Lecture Notes
in Business Information Processing

465

Series Editors

Wil van der Aalst ⓘ
  RWTH Aachen University, Aachen, Germany
John Mylopoulos ⓘ
  University of Trento, Trento, Italy
Sudha Ram ⓘ
  University of Arizona, Tucson, AZ, USA
Michael Rosemann ⓘ
  Queensland University of Technology, Brisbane, QLD, Australia
Clemens Szyperski
  Microsoft Research, Redmond, WA, USA

More information about this series at https://link.springer.com/bookseries/7911

Jacek Maślankowski · Bartosz Marcinkowski ·
Paulo Rupino da Cunha (Eds.)

Digital Transformation

14th PLAIS EuroSymposium
on Digital Transformation, PLAIS EuroSymposium 2022
Sopot, Poland, December 15, 2022
Proceedings

 Springer

Editors
Jacek Maślankowski 🆔
University of Gdansk
Sopot, Poland

Bartosz Marcinkowski 🆔
University of Gdansk
Sopot, Poland

Paulo Rupino da Cunha 🆔
University of Coimbra
Coimbra, Portugal

ISSN 1865-1348   ISSN 1865-1356   (electronic)
Lecture Notes in Business Information Processing
ISBN 978-3-031-23011-0   ISBN 978-3-031-23012-7   (eBook)
https://doi.org/10.1007/978-3-031-23012-7

This Springer imprint is published by the registered company Springer Nature Switzerland AG
The registered company address is: Gewerbestrasse 11, 6330 Cham, Switzerland

In Memoriam of Professor Stanislaw Wrycza (1949–2022) General Chair of EuroSymposia 2007, 2011–2021

Professor Stanisław Wrycza was a General Chair of the EuroSymposium conferences in 2007 and 2011–2021. In his research career, he was the head of the Department of Business Informatics at the University of Gdańsk (from 1991), the initiator and first president of the Polish Society for Information Systems (1995–2000), and president of the Polish Chapter of the Association for Information Systems (from 2006), among many other contributions. He was also recognized as an AIS Distinguished Member for demonstrated commitment to the Association for Information Systems (2021).

He was author and co-author of over 200 scientific publications, including over 40 books. He acted as a member of over 100 Program Committees of international conferences. He supervised over 20 doctoral dissertations and about a thousand master's and bachelor's theses. He organized many conferences, including ECIS 2002, BIR 2008, and numerous events in the EuroSymposia series.

Preface

The PLAIS EuroSymposium 2022 was organized with the leading topic of "Digital Transformation". The papers included in the proceedings are related to the use of machine learning, big data, and the Internet of Things in various applications. Other topics of the proceedings concern the current situation of ICT employees and their creativity via social media channels.

The objective of the PLAIS EuroSymposium 2022 was to discuss the general issues of digital transformation and related topics, as listed on the conference website. The EuroSymposia were initiated by Keng Siau, and previous EuroSymposia were organized by different academic institutions:

- University of Galway, Ireland: 2006
- University of Gdańsk, Poland: 2007
- University of Marburg, Germany: 2008
- University of Gdańsk, Poland: 2011–2021

The papers accepted for presentation at previous Gdańsk EuroSymposia were published in the following proceedings:

- 2nd EuroSymposium 2007: A. Bajaj, S. Wrycza (eds), Systems Analysis and Design for Advanced Modeling Methods: Best Practices, Information Science Reference, IGI Global, Hershey, New York, 2009
- 4th EuroSymposium'2011: S. Wrycza (ed.), Research in Systems Analysis and Design: Models and Methods, LNBIP 93, Springer, Berlin, 2011
- Joint Working Conferences EMMSAD/EuroSymposium 2012 held at CAiSE'12: I. Bider, T. Halpin, J. Krogstie, S. Nurcan, E. Proper, R. Schmidt, P. Soffer, S. Wrycza (eds.), Enterprise, Business-Process and Information Systems Modeling, series: LNBIP 113, Springer, Berlin, 2012
- 6th SIGSAND/PLAIS EuroSymposium'2013: S. Wrycza (ed.), Information Systems: Development, Learning, Security, Series: Lecture Notes in Business Information Processing 161, Springer, Berlin, 2013
- 7th SIGSAND/PLAIS EuroSymposium'2014: S. Wrycza (ed.), Information Systems: Education, Applications, Research, Series: Lecture Notes in Business Information Processing 193, Springer, Berlin, 2014
- 8th SIGSAND/PLAIS EuroSymposium'2015: S. Wrycza (ed.), Information Systems: Development, Applications, Education, Series: Lecture Notes in Business Information Processing 232, Springer, Berlin, 2015
- 9th SIGSAND/PLAIS EuroSymposium'2016: S. Wrycza (ed.), Information Systems: Development, Research, Applications, Education, Series: Lecture Notes in Business Information Processing 264, Springer, Berlin, 2016
- 10th Jubilee SIGSAND/PLAIS EuroSymposium'2017: S. Wrycza, J. Maślankowski (eds), Information Systems: Development, Research,

Applications, Education, Series: Lecture Notes in Business Information Processing 300, Springer, Berlin, 2017

- 11th SIGSAND/PLAIS EuroSymposium'2018: S. Wrycza, J. Maślankowski (eds), Information Systems: Research, Development, Applications, Education, Series: Lecture Notes in Business Information Processing 333, Springer, Berlin, 2018
- 12th SIGSAND/PLAIS EuroSymposium'2019: S. Wrycza, J. Maślankowski (eds), Information Systems: Research, Development, Applications, Education, Series: Lecture Notes in Business Information Processing 359, Springer, Berlin, 2019
- 13th SIGSAND/PLAIS EuroSymposium'2021: S. Wrycza, J. Maślankowski (eds), Information Systems: Research, Development, Applications, Education, Series: Lecture Notes in Business Information Processing 429, Springer, Berlin, 2021

The 14th EuroSymposium, which took place on December 15, 2022, was organized by the Polish Chapter of AIS (PLAIS) and the Department of Business Informatics of the University of Gdańsk, Poland.

The paper submission and reviewing processes were supported by the new EquinOCS system hosted by Springer. Each submission was reviewed by at least two Program Committee members in a double blind manner. According to the review scores, eight papers were accepted for publication in this volume, giving an acceptance rate of 35%. The accepted papers are organized into three parts:

- Artificial Intelligence
- Creativity and Innovations
- Big Data, Internet of Things, and Blockchain Technologies

I would like to thank all authors, reviewers, and Program Committee and Organizing Committee members for giving us the opportunity to engage in high-level discussions on the topics of the conference. With their support, the PLAIS EuroSymposium2022 was a successful event.

This conference was organized in memoriam of Professor Stanisław Wrycza, who was a General Chair for many editions in the EuroSymposia series.

December 2022

Jacek Maślankowski
Bartosz Marcinkowski
Paulo Rupino da Cunha

Organization

General Co-chairs

Jacek Maślankowski   University of Gdańsk, Poland
Bartosz Marcinkowski   University of Gdańsk, Poland
Paulo Rupino da Cunha   University of Coimbra, Portugal

Organizers

The Polish Chapter of Association for Information Systems - PLAIS
Department of Business Informatics at the University of Gdańsk

Patronage

European Research Center for Information Systems (ERCIS)
Committee on Informatics of the Polish Academy of Sciences

International Program Committee

Akhilesh Bajaj	University of Tulsa, USA
João Barata	University of Coimbra, Portugal
Michal Dolezel	Prague University of Economics and Business, Czech Republic
Petr Doucek	Prague University of Economics and Business, Czech Republic
Helena Dudycz	Wroclaw University of Economics, Poland
Jacinto Estima	University of Coimbra, Portugal
Krzysztof Goczyla	Gdansk University of Technology, Poland
Arkadiusz Januszewski	University of Science and Technology in Bydgoszcz, Poland
Piotr Jędrzejowicz	Gdynia Maritime University, Poland
Dorota Jelonek	Czestochowa University of Technology, Poland
Bohdan Jung	Warsaw School of Economics, Poland
Kalinka Kaloyanova	Sofia University, Bulgaria
Marite Kirikova	Riga Technical University, Latvia
Vitaliy Kobets	Kherson State University, Ukraine
Jolanta Kowal	University of Wroclaw, Poland
Tim A. Majchrzak	University of Agder, Norway
Marco de Marco	Università Cattolica del Sacro Cuore, Italy
Paulo Melo	University of Coimbra, Portugal

Ngoc-Thanh Nguyen	Wroclaw University of Science and Technology, Poland
Mieczyslaw L. Owoc	Wroclaw University of Economics, Poland
Nava Pliskin	Ben-Gurion University of the Negev, Israel
Vaclav Repa	Prague University of Economics and Business, Czech Republic
Thomas Schuster	Pforzheim University, Germany
Marcin Sikorski	Gdansk University of Technology, Poland
Piotr Soja	Cracow University of Economics, Poland
Reima Suomi	University of Turku, Finland
Jakub Swacha	University of Szczecin, Poland
Catalin Vrabie	National University, Political Studies and Public Administration, Romania
Samuel Fosso Wamba	Toulouse Business School, France
Janusz Wielki	Technical University of Opole, Poland
Andrew Zaliwski	Whitireia New Zealand, New Zealand
Iryna Zolotaryova	Kharkiv National University of Economics, Ukraine

Organizing Committee

Anna Węsierska	University of Gdańsk, Poland
Dorota Buchnowska	University of Gdańsk, Poland
Bartłomiej Gawin	University of Gdańsk, Poland
Przemysław Jatkiewicz	University of Gdańsk, Poland
Dariusz Kralewski	University of Gdańsk, Poland
Jacek Maślankowski	University of Gdańsk, Poland
Patrycja Krauze-Maślankowska	University of Gdańsk, Poland
Michał Kuciapski	University of Gdańsk, Poland
Bartosz Marcinkowski	University of Gdańsk, Poland
Monika Woźniak	University of Gdańsk, Poland

Contents

Artificial Intelligence

Artificial Intelligence Promises to Public Organizations and Smart Cities

Catalin Vrabie(⊠)

National University of Political Studies and Public Administration, Bucharest, Romania
catalin.vrabie@snspa.ro

Abstract. Artificial Intelligence (AI) is influencing almost every area of modern life from entertainment, commerce, and healthcare to organizations internal processes operations. Netflix knows what movies and series people want to watch, Amazon knows when and where people like to shop, and Google knows what users are looking for. All this information may be utilized to create very comprehensive personal profiles, which can be useful not just for behavioral analysis and targeting, but also for forecasting economic trends, political changes and for understanding people's attitude towards different topics. There is a lot of hope that AI could lead to significant breakthroughs in all aspects of life. It can help in organizing public administration institutions' internal processes as well as urban development and/or transformation. In this article, the author considers both of those directions similarly important to citizens and provides a list of political, economic and administrative benefits in using AI applications in public organizations (at local, regional or national levels). Several ethical considerations regarding the use of AI and its potential impact on the current labor market are included in the discussion section.

Keywords: AI · Governance · Cities

1 Introduction

For many scholars, modern thinking about intelligent systems started when Alan Turing wrote his famous article entitled 'Computing Machinery and Intelligence' [1]. Since then, computing had improved to such an extent that today we have difficulties in pointing out live events that are not somehow assisted by machines. Concepts such as Machine Learning (ML), Natural Language Processing (NLP) and Robotics are no longer bound to the limits of the scientific world. The Big Four – Amazon, Apple, Facebook and Google are using AI in most of their operations; other business operators, not directly linked to the technological market, are already vastly using AI to smoothen their production cycles or gain more efficiency in their management processes [2, 3].

Recently, governments and other public sector organizations started to leverage Robotic Process Automation software to handle repetitive tasks, decrease human errors and improve compliance [4]. Computer vision and video is used more and more to identify misbehaviors and supervise traffic [5]. Sensors and Internet of Things (IoT) devices

J. Maślankowski et al. (Eds.): PLAIS EuroSymposium 2022, LNBIP 465, pp. 3–14, 2022.
https://doi.org/10.1007/978-3-031-23012-7_1

are placed all over the cities to collect data [6]. Behind those operations AI software is the tool in use.

The demand for public services is rising continually and cities are facing labor shortages [7]. Public organizations are struggling to answer the growing expectations of their clients promptly and effectively, just as Amazon, Apple or other companies with incredible R&D sections engaged in technical innovations do. Advances in wireless technology and smartphones have opened the door to on-demand services via different applications and platforms, as well as to a new type of delivery via remote interactions, accessible anywhere and at any time [8, 9]. Such services are useful since they cut expenses and avoid unnecessary spending.

The value of AI-powered tools in smart cities and governments technology is recognized by the whole ecosystem. AI is thought to have the potential to improve any process in internal operations and service delivery [10] while cost reductions associated to AI-applications for the public sector can be a major motivator for its deployment. The shift from a reactive to a proactive approach to delivering public services is undergoing.

The research question from which the author started in writing this article is *knowing the potential value of AI, how Public Administration and cities could benefit from it?* In order to answer it, the author will firstly present the AI tools as they are found in the literature and other research documents and initiatives and then will be able, extracting the benefits, to propose AI solutions for public institutions and cities. There might be however, discussions for each of them, mostly because these technologies are just beginning to arise in the Public Administration field and some were not yet implemented, but their functionalities, technical and also economic benefits were tested in private environments. The article does not intend to confirm the effectiveness of the proposed solutions but only to provide them to readers (some could be policy makers) in order to enlarge their views and, maybe, to start adding them on strategic development city, or institutional, plan.

1.1 Artificial Intelligence Tools

The last decade witnessed several technical advancements in the fields of AI and data science. Even though AI research for numerous applications has a rather long-standing tradition, the present wave of AI hype is unlike any other. Rapid development of AI tools and technologies (applicable to public sector organizations, cities, and governments altogether), has been enabled by a perfect mix of improved computer processing power, larger data collection data banks, and a vast AI skill pool. This results in a significant change regarding AI technology acceptance and influence on society.

Many data and computation-based technologies have grown at an exponential rate. Moore's rule [11], which explains the exponential development in the performance of computer processors, is the most well-known example. Many consumer-oriented applications have seen similar rapid growth by providing low-cost services. The digitalization of public data might result in a similar development pattern in public organizations, cities and governments as computing those data become less expensive and electronic data collecting platforms become more common. Even though these segments of public life appear minor at first, exponential development will eventually take over. Humans tend to exaggerate the influence of technology in the short term (e.g., one year) while underestimating it on the long-term (e.g., ten years) [12].

IBM Watson [13] and Google's Deep Mind [14] are two businesses that are leaders in this field. They have demonstrated that AI can outperform humans in a variety of jobs and activities, such as chess, Go, and other games. Many smart cities applications could make use of IBM Watson and Google's Deep Mind.

1.2 Main Fields of AI [15]

- Machine Learning (ML). This field deals with computer programs that try to learn from experience in order to predict, model or try to understand data. Some of the most important techniques of ML are: Unsupervised learning – when computers learn by themselves having access to data, Supervised learning – when humans, by using examples, are providing computers right answers so they can learn faster and Reinforcement Learning – when computers perform tasks that are changing the structure of nature/reality and they get some sort of rewards/penalties after the output is measured. However, the difference between unsupervised, supervised and reinforcement learning is in the way (i.e., the techniques) the learning process occurs, and also in the kind of tasks they perform.
- Natural Language Processing (NLP). That part of AI is focused in getting computers to understand and generate unrestricted natural languages with the flexibility and fluence.
- Robotics and Robotic/Intelligent Process Automation (RPA). RPA with its Intelligent Process Automation (IPA) extension subfield is used to automate repetitive tasks both in the back office and front office that require human intervention. 'Intelligence' in this regard means that computers are not only automating tasks but also understand the content so to adjust the actions according to it (Fig. 1).

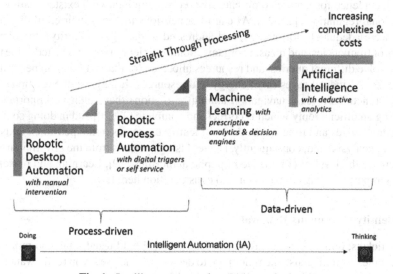

Fig. 1. Intelligent Automation (IA) complexity [1]

2   Artificial Intelligence in Public Organizations

2.1   Petitioning

Closely linked to the freedom of expression and thought, as well as to the concept of political accountability, the right to petition lies at the core of local democracies [18, 19]. While municipalities strive to become more participatory and open towards their citizens, they create facilities to interact via the internet. E-petitioning, for instance, is one of those easy-to-handle tools presently in use across the world, in both local and central governments. It is true that, nowadays, many public organizations use social media networks to interact with their communities and many complaints, comments or suggestions are addressed by citizens or legal persons via Twitter, Facebook, and others. AI can help in these instances too, by providing short answers or prompt the authority to answer. Still, in most cases, discussing over the social media has no legal consequences in many administrative courts (e.g., if a citizen raises a complaint, it is not seen as an official one, as the case with petitions). It is for this reason that I consider official petitioning an AI problem-solving solution [20–22].

Today, citizens expect their public authorities to be prompt, excellence-driven and effective; and as levels of trust in governments and their services continues to drop throughout the world [23], better infrastructure, more qualitative services, and an adaptive leadership are loudly demanded. With the needs of the many increasing, public budgets face serious constrains: the literature on public management advises in fact that long delays in presenting effective solutions, poor skilled personnel, and overall poor administrative capacities may have smoothen the path for AI applications in generating and sustaining good governance.

In the specific case of the e-petitioning, AI can advise the triage, offer systematic and automatic replies to some queries and decide which petitions require further analysis from a specialized department. It could also assist the decision making by providing, on demand, evidence for a more comprehensive reply, compliant with existent national or international regulations [24–26]. AI can filter petitions to verify their eligibility (e.g., if they were adequately formulated, and addressed to the right authority), to compare subjects or frequencies, and measure organizational efficiency (e.g., translated in the time spent to solve different queries, and resources allocated for that purpose). In performing these tasks, AI indeed saves time, energy and resources. It may use the 'compare and comply' functions to ease navigation among regulations that might be important for delivering an official reply which is correct and complete [27, 28] and in doing so it will limit redundancies and time waste. AI can identify urgencies in the petitions' texts (by sentiment analysis) and, consequently, trigger faster reactions from the government side and increase the level of confidence in public authorities [29]. Learning and reasoning are also important things to be taken into consideration here [30].

2.2   Identity Documents Renewal

Buying tickets, online, to your favorite artist's show, should usually take no more than three minutes, counting also the time spent to decide in which seat you really want to sit. Making a doctor's appointment online, should also be possible in less than five minutes (considering you will need to input more information about your medical history, and decide which day and what interval better suits your agenda). Buying your groceries online would also take you no more than 15 min if no previous list of groceries exists in the application and you take your time to browse different products to decide which price/quality favors your tastes. It is no wonder then that people expect that a one-page document, stating your identity, should be easy (under five minutes) to retrieve from an official governmental portal. However, to renew your identity documents in many of the European states still takes 10 days on average and usually involves a face-to-face interaction with at least one civil servant. This may be because there are no e-solutions available, or, if there, redundancy seriously affects their overall effectiveness. RPA could evaluate and reduce such redundancies, by analyzing submissions, verifying and certifying personal information, approving or rerouting applications for review [31]; generally, that would mean they could assist with (but not be limited at): driver's license applications and renewals; passport applications and renewals; national ID application and renewals.

2.3 Public Procurement

In these turbulent times, when "wicked problems" [2] need governmental attention more than ever, reducing the waste of time and resources allocated to basic, routine tasks becomes essential. E-procurement systems can get a lot of help from RPA/IPA systems. If applied, civil servants will only fill in a form with details about the products/services they require and the system will complete the following, necessary steps, deal with deadlines and existent regulations. In return, procurement specialists will be able to focus on maintaining vendor relationships, forecast interventions and discuss strategic alternatives.

In a recent document UiPath [3], described and proposed to areas of RPA implementation: help business environment recover (after crises) and grow (1) and streamline and modernize government (2). In regard to the second area, General Services Administration (GSA) using Document Understanding have automated the process of going through, and verifying, thousands of documents (comparing them with legislation and contracts) in minutes (tasks that were previously handled by humans in hours).

2.4 Parking Permits Renewal

Residential areas could also get help from an RPA system. In order to renew their parking permit, citizens may use a dedicated application which may extend the yearly validity of their document. In most cases, as of today, even if the system is digitalised it is not automated and by that it means that most of the task – being handled by humans in back-office operations, are repetitive and subject of errors. Computers can easily jump into the scene and replace human labour. For example, the system will search if there are any changes by comparing the newly received documents with previous records stored in local databases (i.e., address, plate numbers, car type) and if the case, will make the necessary inquiries, fill in the new (or missing information) or reroute the submission to an operator for further references.

The author has no information about such a system being used for this purpose but similar applications are working in private sector in various domains such as accounting – by the use of data entry when paying invoices [4], and even insurance – when artificial intelligence bots are paying the claims with less or no human interaction with the customer [5].

2.5 Report Generation

Cutting the red tape has been one of the main objectives of public reforms in the past decades all over the world. AI applications can assist with this task, by offering solutions for generating reports on budgets, spending, operations, citizen requests, grants and other specific tasks. In return these may conduct to administrative simplification and more accountability from the public sector's side.

Since a decade ago some companies developed AI techniques based on NLP who are able to turn data into numbers – Automated Storytelling [6], to create news articles from just raw data or, as Kris Hammond from Narrative Science said "to humanize data. It is to be a communication bridge between the numbers and the knowing".

Data for these reports are obtained from the organizations' legacy systems. Optical Character Recognition (OCR) features are used by RPA bots to read text in photos and documents, extract important data, and create reports [7].

2.6 Reception "Officers"

During crises situations or simply on daily basis, to assist the vulnerable or general population, hardware robots, such as Pepper or Relay, could be used by public organisations at local or national levels (e.g., in city halls, Ombudsman Offices, etc.) to assist citizens in identifying the right desks to address their inquiries or promptly solve some of their grievances. In the private sector similar application of robots are to be found in hospitality industry such as hotels and other places who needs humans to help directing and routing clients and customers.

At the present time, many public institutions do use fixed machines able to issue Walk-in Tickets [8, 9]. The next step will be using reception "officers" that could be programmed to recognize citizens based on computer vision (face recognition) and address filter questions in order to provide the best available advice [10, 11] while, as a back-office task, they are already preparing the files for the desk officer assigned to take care of the case. This will ease the interaction between the citizen and the public servant as well as will provide a friendly environment for dialog.

2.7 Public Sentiment Analysis

In a culture still dependent on a new public management (NPM) rhetoric, focused on entrepreneurial solutions and market-based tools for boosting public efficiency, RPA bots can be used to gather information from Internet platforms on public perceptions on various government agencies, services, or officials. Bots can scrape people's comments or ratings of government services and facilities to study public mood and improve services appropriately [12–14].

3   Artificial Intelligence in Cities

3.1   Traffic Management

Mobility is a very important aspect in any city across the world; With the growing trend of urbanization, people tend to make their agendas to avoid traffic congestions, transporters nudge their clients with discounts to stimulate the use of their services outside the rush hour, public organisations reschedule some of their public hours in order to meet the demands of their communities and automobile manufacturers, as well as other service providers search for solutions to make the "wait in traffic" more pleasant for the drivers [15, 16].

AI can be of help: it can collect data from different drivers and predict behaviours, thus adjusting the traffic flow, and improving the overall mobility of all commuters. One can argue that apps like Waze and/or Google maps are already doing it, so AI are redundant in this case; nothing further from the truth. Such applications are only foreseeing traffic conditions on a short (or very short) time period [17, 18]. However, AI could run on Google maps platform collect traffic data for a year or more (e.g. to learn of traffic conditions in all seasons, etc.) and predict drivers' behaviours in different contexts such as: weather (e.g. how will a light shower or a heavy snow affect the traffic), time (e.g. daytime, rush hour, etc.), holidays (e.g. weekdays versus weekend), school calendar (e.g. school activities usually prompt parents to use their personal cars more than in weekends), and consequently provide drivers with more efficient routes.

AI can communicate with drivers by NLP techniques listening to their voices [19] and eventually generating text over data [7] and read it [20] to provide more information to them in regard to the route they need to drive through.

Fixed traffic sensors (cameras) can also monitor the road activity and send the data to servers thus adjusting the inputs from the drivers' mobile devices and enriching the system capabilities – in the situation where public utility cars/security/ambulances are forced to intervene, those vehicles will mainly use this network of sensors/cameras to prevent cases of hacking such as the one registered in Berlin where an 'artist uses 99 phones to trick Google into traffic jam alert' [21].

Next step will be to synchronize the data with the traffic light system and, based on Internet of Thing (IoT) devices, it will be able to control the colour changing and to indicate to drivers by the use of arrows different routes to avoid congested junctions. The system should be able to predict based on different hours, days or events what are the most congested areas, therefore it could indicate optimal routes, estimate timings etc. It should also be able to anticipate collisions due to traffic/weather or other variables and by that to avoid it by redirecting drivers through their mobile apps or by traffic lights signals (Fig. 2).

In some cases, the data needs to be presented in narrative reports: public organisations, together with the involved stakeholders, should be able to do that by NLP techniques [7]. This might be important for mass-media, for legal initiatives or explanations (e.g., in case of a system failure) and so on. Those side aspects of implementing AI in a traffic management system should be taken into consideration as well.

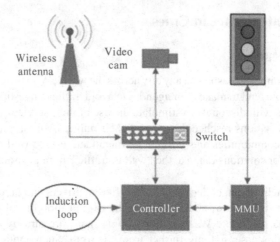

Fig. 2. Control and monitoring equipment at a typical road intersection [22]

3.2 Parking Management

AI could also assist in mapping parking places and, by use of ML techniques on data collected over time, the system will be more efficient in using the available parking places – here seen as resources. The system will be able to point out to drivers (and even reserve on their behalf) a designated slot where they are able to temporarily park their car in the proximity of the requested area. Assuming that all parking slots are somehow private – so, when the owner is coming back home, he/she needs to park his/her car on it, the system, based on previous collected data, will also predict the availability of each and provide this info to drivers. However, if by accident, the owner needs his/her place earlier then presented by statistics, the system will contact the driver and provide another option in the near vicinity.

The main beneficiary of this implementation will be, of course, the citizen, the individual. He/she will spend less time in traffic, will reduce fuel consumption and will be less stressed. However, public organisations, as well as communities, will also benefit from it. Less pollution and reduced stress bring along a healthy population. Less or no traffic congestions will release the pressure otherwise placed on police forces dealing with traffic issues daily. The likelihood of a decline in car incidents/accidents is to be expected, and by that, indirectly, general benefits for the health system are to be envisaged. An impact on the car insurance business is also foreseeable.

3.3 Waste Management

AI can be useful in the waste management services as well. Sensor equipped trash bins can send a signal to the garbage collector company when it is 80% full [23, 24]. By reinforcement learning, the system may optimize the trash collecting vehicles routes to efficiently collect the waste and not allow overflowing.

The adoption of AI in waste management could be a solution since path improvement and analytics data use led to saving 30% of waste collection costs [25], in addition to reducing fleet fuel consumption, carbon emissions and traffic congestion.

Bergen, Norway's second-largest city, offers one of the best, if not the best, waste management system, which is capable of handling waste straight from the households, using airflow alone throughout an underground pipe system without requiring on-street waste containers [26] (however, those being still in use by pedestrians). Here, once the incoming waste is getting to the sorting belt, scanners equipped with sensor units connected to AI applications are able to detect its characteristics and robotic arms or conveyors are carrying waste to the correct container helping the entire recycling process.

3.4 Autonomous Vehicles

Introducing level 4 autonomous vehicles [27] firstly on dedicated routes for pupils going to school (by minibuses like Olli which is equipped with robot navigation technologies [28]) and extend it to other public transportation routes, can also be an important step forward in ensuring the good governance of cities. Olli is easily spotted in traffic by drivers, it uses only its dedicated routes and sends data to the system communicating with the mobile devices of the nearby drivers. In doing so, it provides a safe journey to children.

4 Possible Consequences of AI Implementation

4.1 Job Shortage "Risks"

AI is not easy to design, nor to operate. To properly function, teams of engineers and dedicated personnel are needed. Maintenance operations need to be regularly conducted both at the street level – taking care of the IoT devices and cameras as well as at the servers' data centre. Most probably public organisations and cities will need to outsource some processes to specialised companies with experience in specific areas of technology [29]. However, dedicated personnel need also to be trained and used by governments as well [30]. A dedicated category of employees should supervise the system to teach it better – for example, in regard with traffic management, when a congestion is spotted, generated by a collision, or information from a construction site is being added, they need to let the application learn to be able to redirect the traffic by simply adjusting the parameters. Similarly in regard with parking management, when they observe that specific parking areas (e.g., in residential areas) are becoming empty after locals are going to work during the mornings, they can teach the system about that too. Their inputs, regarding all those situations where the supervised and active learning are in place, are valuable and, to keep the system up and running and always improving, lot of this labour need to be hired.

Outsourcing means bringing multiple stakeholders to the same table. IT companies will be needed to handle the software needs and updates, and transport and service companies for hardware maintenance. Most probably an update to the actual traffic legislation will be needed, therefore different public institutions need also to participate with their knowhow.

4.2　Lack of Trust and Ethical Concerns

Governments need to adopt a data strategy that focuses on AI. Public entities need to understand the collected data, to extract value from it, and to provide value to citizens [31]. However, technology is not perfect, and systems do make mistakes. The major risk here is for users to lose trust and stop using AI solutions. That might be overcome by accepting the made-mistakes and, if necessary, by paying compensations to the affected ones.

Other reasons that might keep users away from using AI solutions may concern privacy: one way or another AI identifies users, so, in the traffic management case, it learns of their GPS location. This thing might be sensitive considering personal beliefs and ideologies. To overcome this, officials can offer assurances that the data collected is fully anonymized and only used for improving the traffic.

4.3　Lifelong Learning for AI-Literacy

To fully appreciate the benefits AI solutions can bring to our daily lives and to the overall welfare of our communities, lifelong learning opportunities should be created. Raising awareness on what AI means and can do might benefit not only the younger members of our societies, but the elder as well. One of the main challenges I believe we need to face is to become AI-literate as soon as possible.

5　Discussions and Conclusions

Today, AI systems are driving advances in medical technology, transportation and more. Governments are adopting AI to improve administration, disaster response and security. Existing approaches, policies and further down legislation need to be re-evaluated to ensure that AI delivers on its potential for truly positive impact. Before this potential can be realized, however, a number of challenges need to be addressed.

- Inconsistent performance;
- Safety, especially if used in sensitive context like self-driving vehicles;
- Responsible use;
- Impact on society, including employment, social and civic engagement.

While is difficult to fully solve these challenges, being aware of them and acting accordingly will certainly reduce the impact. Moreover, this should be the starting point for creating a safe AI environment that is also beneficial to all users.

The opportunities associated with AI are immense. Because of this, there is a need to rethink how AI systems are regulated, and to rethink the institutional structures.

Also, an appropriate information campaign, education of residents and a wide-ranging dialogue with stakeholders and enterprises that would decide to implement this type of technology in urban space may determine the success of AI implementation in smart cities. However, further studies need to be undertaken on this topic. The author is actively engaged in building up cities development strategic plans as well as scanning the deployment of smart technologies in urban environments in Romania willing to promote the benefits of AI among policy makers.

References

1. CFB Bots: The Difference between Robotic Process Automation and Artificial Intelligence. CFB Bots (2018). https://cfb-bots.medium.com/the-difference-between-robotic-process-aut omation-and-artificial-intelligence-4a71b4834788. Accessed 6 May 2022
2. Pollitt, C., Bouckaert, G.: Public Management Reform: A Comparative Analysis—Into the Age of Austerity. Oxford University Press, Oxford (2017)
3. UiPath: Unleashing the Power of RPA Across the U.S. Economy. UiPath (2022)
4. Zhang, C.A., Issa, H., Rozario, A., Søgaard, J.S.: Robotic process automation (RPA) implementation case studies in accounting: a beginning to end perspective. Forthcoming in Accounting Horizons (2022)
5. Lemonade: Lemonade. Lemonade. https://www.lemonade.com/fr/en/?f=1. Accessed 26 Oct 2022
6. Hammond, K.: Interviewee, Automated Storytelling: Kris Hammond for the Future of StoryTelling [Interview] (2012)
7. Hammond, K.: Automated Storytelling. FoST (2012). https://futureofstorytelling.org/video/ kris-hammond-automated-storytelling. Accessed 6 May 2022
8. Sintezis; S-ticket sistem eliberare numere de ordine. Sintezis. https://www.sintezis.ro/indust rie/s-ticket-sistem-eliberare-numere-de-ordine/. Accessed 27 Oct 2022
9. Q-net: Queue Management System. Q-net.https://q-net.pro/. Accessed 27 Oct 2022
10. Relay Delivers Hospitality. Savioke (2022). https://www.relayrobotics.com/. Accessed 6 May 2022
11. Humanizing technologies: Emotional appeal to customers through the humanoid robot Pepper. Humanizing technologies (2022). https://humanizing.com/en/pepper-robot-humanoid-robot-by-softbank-robotics-for-retail-fairs-receptionist-showrooms-happiness-hero/?utm_term=pepper%20robot&utm_campaign=Europe/Search/Pepper&utm_source=adwords&utm_medium=ppc&hsa_acc=5199685425&hsa_cam=10873. Accessed 6 May 2022
12. Barnhart, B.: The importance of social media sentiment analysis (and how to conduct it). Sprout Social (27 March 2019). https://sproutsocial.com/insights/social-media-sentiment-ana lysis/. Accessed 6 May 2022
13. Dabhade, V.: Conducting Social Media Sentiment Analysis: A Working Example. Express Analytics (24 May 2021). https://www.expressanalytics.com/blog/social-media-sentiment-analysis/. Accessed 6 May 2022
14. Jindal, K., Aron, R.: A systematic study of sentiment analysis for social media data. In: Materials Today: Proceedings (2021)
15. Vrabie, C., Dumitrascu, E.: Smart Cities de la idee la implementare. Universul Academic, Bucharest (2018)
16. Ion, D.G.: Solving the traffic issue. SCRD 1(1), 65–72 (2017)
17. Petreanu, Y.: Under the Hood: Real-time ETA and How Waze Knows You're on the Fastest Route. Waze (27 August 2020). https://medium.com/waze/under-the-hood-real-time-eta-and-how-waze-knows-youre-on-the-fastest-route-78d63c158b90. Accessed 6 May 2020
18. Lau, J.: Google Maps 101: How AI helps predict traffic and determine routes. Google (3 September 2020). https://blog.google/products/maps/google-maps-101-how-ai-helps-pre dict-traffic-and-determine-routes/. Accessed 6 May 2022
19. Appen: An Introduction to Audio, Speech, and Language Processing. Appen (22 April 2021). https://appen.com/blog/an-introduction-to-audio-speech-and-language-proces sing/. Accessed 6 May 2022
20. Google: Put Text-to-Speech into action. Google (2022). https://cloud.google.com/text-to-speech. Accessed 6 May 2022

21. Weckert, S.: Berlin artist uses 99 phones to trick Google into traffic jam alert. The Guardian (3 February 2020). https://www.theguardian.com/technology/2020/feb/03/berlin-artist-uses-99-phones-trick-google-maps-traffic-jam-alert. Accessed 6 May 2022

22. Ghena, B., Beyer, W., Hillaker, A., Pevarnek, J., Halderman, J.: Green lights forever: analyzing the security of traffic infrastructure. In: 8th USENIX Workshop on Offensive Technologies (WOOT 2014) (2014)

23. Nord Sense: Smart Bin Sensors. Nord Sense (2022). https://nordsense.com/smart-bin-sensors/. Accessed 6 May 2022

24. Wen, X.-Y.: City intelligent life: a case study on Shenzhen city intelligent classification of domestic waste. SCRD 5(1), 27–30 (2021)

25. Dhayyat, S.: Smart waste management using artificial intelligence. Smart City J. https://www.thesmartcityjournal.com/en/green-new-deal/smart-waste-management-using-artificial-intelligence. Accessed 27 Oct 2022

26. Infrastructure Intelligence: World's largest automated vacuum waste collection system set for Bergen (2016). http://www.infrastructure-intelligence.com/article/nov-2016/world%E2%80%99s-largest-automated-vacuum-waste-collection-system-set-bergen. Accessed 5 May 2022

27. SAE: Taxonomy and Definitions for Terms Related to Driving Automation Systems for On-Road Motor Vehicles. Society of Automotive Engineers (SAE) (2018)

28. Marineterrein Amsterdam: Testing autonomous transport (24 October 2019). https://www.marineterrein.nl/en/project/meet-olli-the-self-driving-minibus/. Accessed 4 May 2022

29. Olusegun Fayomi, J., Abdulqadir Sani, Z.: Strategies for transforming the traditional workplace into a virtual workplace in smart cities. SCRD 6(1), 35–54 (2022)

30. Virtosu, I., Li, C.: Bundling and tying in smart living. SCRD 6(2), 97–110 (2022)

31. Schachtner, C.: Wise governance – elements of the digital strategies of municipalities. SCRD 6(2), 23–29 (2022)

Sustainable Robo-Advisor Bot and Investment Advice-Taking Behavior

Vitaliy Kobets[1(✉)] ⓘ, Oleksandr Petrov[1] ⓘ, and Svitlana Koval[2] ⓘ

[1] Kherson State University, 27, Universitetska St., Kherson 73003, Ukraine
vkobets@kse.org.ua
[2] Kherson State Agrarian and Economic University, 23, Stritenskaya St., Kherson 73006, Ukraine

Abstract. One of the main reasons why people use robo-advisors is that the traditional financial instruments (e. g. deposits, bonds) will promise zero returns in the near future. Robo-advisors using built-in algorithms to determine the assets of investment portfolios for short-run and long-run periods will be one of the promising options for the investor to obtain income in order to achieve his/her goals. Live Trading bots will be able to provide substantial passive in-come for investors, that is an attractive alternative, compared to the advice of traditional human advisors. The goal of this paper is to develop a robo-advisor bot to make investment decisions in order to choose the best financial instruments considering risk-return criterion using different investment strategies. The paper deals with the models of robo-advisor bot for different risk attitudes of investors. Each investor can choose among different investment strategies, such as buy-and-hold strategy, moving average strategy, relative strength index strategy, support and resistance strategy with different performance measures. All strategies dealing with risk-return criterion for precious metals demonstrate the greatest efficiency for risk-averse investors. RSI (relative strength index), buy-and-hold strategies are also effective for Netflix shares. Oil and cryptocurrencies are most appropriate for different strategies of risk-seeking investors. Tesla stock is the most appropriate for risk-neutral investors under definite period.

Keywords: Robo-advisor bot · Risk-return criterion · Investment strategy · Financial instruments

1 Introduction

H. Markowitz's Modern Portfolio Theory (MPT) is based on two main factors: risk and expected returns. In this case, an investor chooses a portfolio with the highest returns and the slightest danger. The investment goal is to get more profit with less risk [1]. Building a portfolio an investor can act in two possible situations: complete uncertainty (an investor can not determine scenario probabilities) and conditions of risk (probabilities can be determined). Nowadays, to clarify the model we can add one additional factor – risk tolerance or risk aversion which is influenced by various factors and described by the Arrow-Pratt coefficient. It is valid only if an investor behaves rationally: he/she can

J. Maślankowski et al. (Eds.): PLAIS EuroSymposium 2022, LNBIP 465, pp. 15–35, 2022.
https://doi.org/10.1007/978-3-031-23012-7_2

calculate different scenarios, identify the utility, maximize benefits, always choose the best option among the alternatives.

However, even experienced investors often make controversial and sometimes wrong financial decisions, for example an undiversified portfolio or a risk concentration. C. Frydman and C.F. Camerer associate it with low level of financial literacy and managed funds popularity [2]. Moreover, we tend to see the main problem of investors' mistakes in cognitive constraints. According to D. Kahneman's research, the decisions made by economic agents usually differ from those made with the "homo economicus" model [3]. J.Y. Campbell insists that "households do not save and invest according to the normative models" [4]. Consequently, they "typically have underdiversified stock holdings and low retirement savings rates" [2]. De Bondt says that people forget the basic investment principles and laws when investing; they rely on intuition and other factors but not quantitative measures [5].

It is true not only for the private investors but for the experienced, financially literate managers as well: "even the top business managers, who are usually highly educated, make decisions affected by overconfidence and personal experience" [2]. And what is more, "even Markowitz, the creator of MPT, did not use MPT in his own portfolio choice" but he simply created a 50/50 mix of stocks and bonds [6]. The issues dealing with growing data volumes, lack of understanding information culture, low level of economic agents' financial literacy, cognitive constraints will intensify. Moreover, E. Bikas claims investing in financial markets is becoming more popular [7]. This requires the use of automated financial and investment decision-making tools.

Now there is a large number of software solutions for robo-advisors. Chatbot service is actively used and has already been implemented in many sectors for customer support. Let us consider an example of such a bot Erica (Fig. 1), offered by Bank of America. The bot was introduced in the end of 2018 and at the beginning of 2019, it had 6.3 million users and 16.5 million applications. This solution is built into the bank's mobile application; it involves the functions: sending notifications about banking alerts, identifying available opportunities to reduce costs, i.e. counseling customers and giving advice how to save money, notifying about credit rating changes and simplifying bill payments.

Further, we consider the Ally Assist bot (Fig. 2) launched in 2015 by Ally Bank. Ally Assist is a chatbot built into the Ally Mobile app that you can interact with using voice commands and text messages. The functions of the bot involve making payments, money transfers, P2P transactions and deposits. Also, the user can get information about the bank account or transaction history. Using Machine Learning (ML), Ally Assist can predict a user's needs through account and transaction analysis in order to provide relevant messages with recommendations. In addition, the bot uses Natural Language Processing technologies to answer frequently asked questions.

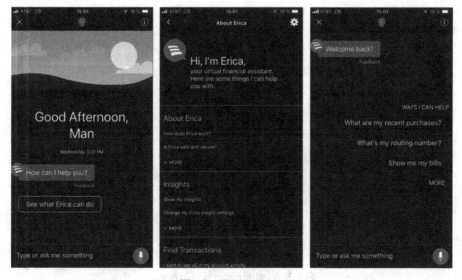

Fig. 1. Chatbot Erica.

Fig. 2. Chatbot Ally assist.

Robo-advisors can be considered as an application. They provide affordable opportunities to invest and manage money in traditional models, allowing more people to save, invest and grow their capital to achieve financial goals. There are many examples of robo-advisors, but the best of them is Wealthfront (Fig. 3).

Fig. 3. Wealthfront robo-advisor.

Wealthfront offers an exceptional portfolio management experience and remains a digital, automated financial investment solution with an extensive suite of portfolio management tools and a wide variety of financial products.

As far as we know, there is high demand for robo-advice and possibility for investor to choose among different opportunities, but there are no platform or application to meet these requirements.

The **goal** of this paper is to develop a robo-advisor bot to make investment decisions in order to choose the best financial instruments considering risk-return ratio using different investment strategies.

The paper has the following structure: Sect. 2 considers the related works to the study. Sect. 3 presents the models of robo-advisor for the investor's different goals. Sect. 4 considers strategies of robo-advisor bot for different financial instruments. The last part concludes the paper.

2   Related Research

We use the economic, statistical and analog methods, the expert views to assess the financial risk. The economic and statistical methods assess financial risk using the following indicators: the average value of the investor's profit of random variables (risk factor); dispersion; standard deviation of a profit; semi-standard deviation; the coefficient of variation; probability distribution of a profit. The density function of a normal probability distribution allows us to calculate the probability of making a profit. Value at Risk (VaR) or "investment risk" is an integral measure of risk that can compare the risk of different investment portfolios and different financial instruments. VaR shows a confidence of x% (with a probability of x%) that the investor's losses will not exceed y monetary units over the next n days. In this statement, the value of y is unknown and is

VaR, which is a function of two parameters: the time horizon n and the confidence level x [8].

Characteristics of private and institutional investors' impact on the probability of using sustainable robo-advisors, and probability of use is 1.53 times higher among young and experienced investors. RAs use the mathematical algorithms and artificial intelligence in order to advise clients and reiterate human service. 'While people can be influenced by emotions, so it leads to wrong investment decisions, RAs claim to be bias free' [9]. COVID-19 pandemic was the first test for RA and this showed investors 'that online services are vital for investment purposes' using exchange traded funds (ETF) mostly.

The EU policy requirements to classify sustainable and non-sustainable investments and competition between RAs tend to ethical investing using the green investment in order to fight climate changes.

The authors [10] demonstrated that the investor's sustainable consumption 'transforms into a higher probability of choosing the portfolio according to a sustainable investment strategy', 'the sustainable consumers are more likely to select a green robo advisor that offers the sustainable investment strategies even though this requires higher management expenses' and 'socially responsible investment is an opportunity for investors to invest in accordance with their personal values'. At the same time, diversification of green investments portfolio is decreased and creates higher costs of RA.

According to the return perspective, it is important to distinguish conventional (not considering the impact of investments on the environment) or sustainable (e.g., return from green investments) investment decisions. It means a trade-off between social and financial returns depends also on the investors' utility preferences. It can be done by implementing smart algorithms and artificial intelligence tools.

We can define 3 different types of RA (Fig. 4) having distinguishing features (Table 1) [9].

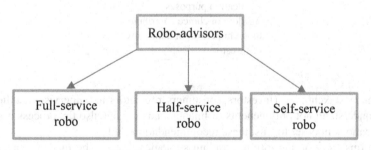

Fig. 4. Types of robo-advisors.

Investor's characteristics (male, age, education, cost awareness, ecological aspects) play a significant role for the probability of investing via a sustainable robo-advisory service, where mostly older persons prefer to use a sustainable robo-advisor [9].

RA provides with advanced estimation techniques employing artificial intelligence, but it still has the characteristic of being understandable. The investors who require more RA's assistance are less likely to use these services due to lack of trust. And for that

Table 1. Types, goals and characteristics of RA.

Types of RA	Characteristics of RA	Goal
Full-service robo (robo-managers)	Monitoring the robo-advisor's business for investor protection Sequential recommendations to meet the complex and dynamic investment requirements Reallocations in investment portfolios	Automated portfolio management service (automated decision making) in long-run period (more than 10 years)
Half-service robo (traditional of hybrid)	Always require the investor's consent to execute orders; The percentage weight can vary; the offensive share is stated by stocks whereas the defensive aspects refer to conservative investments such as bonds Only a portfolio suggestion is after an initial conversation with a human financial advisor	The provision of investment proposals robo-advisor does not hold a mandate to autonomously execute orders but rather acts as an investment intermediary in middle period (5–10 years)
Self-service robo	Self-service-robo neither executes an order, nor opens a securities account An information source is used to gain investment proposals To gain additional information or investment proposals for verification purposes freedom of choice contributes to the high level of autonomy of investors	To provide information to an investor thereby assisting in the decision-making process to independently manage the portfolio in short-run period (1–3 years)

reason, many stakeholders (investors, consumers, regulators and supervisory authorities) need to understand not only benefits of financial advice, but also the process itself and its implications in order to trust these recommendations [11].

RA firms 'rely on algorithmic trading strategies that can be made automatically, without hardly any human decisions in the process' [12]. There are qualitative and quantitative differences between RA within and across different countries (the USA, the UK, the EU). 'The aim of the qualitative analysis is to use the predefined criteria to gain a better understanding of the heterogeneity of the RAs approach to portfolio selection and subsequent portfolio management' [12]. The qualitative criteria in the fields of law, economics, and computer science are described in Table 2.

Table 2. Robo-advisors qualitative criteria.

Qualitative criteria	Description of criteria	Comments (examples)
Type of RA	Information about unique nature of business model of RA and its algorithms	White paper of RA
Amount of assets under management (AUM)	A *time series database* has to be generated	Markets of the USA, the UK, Germany from the largest till the smallest AUM
Investment portfolio data availability	Data retrieval and processing about financial instruments (e.g., shares, ETF, cryptocurrencies, real estate, precious metals, currencies, commodities)	Open data: https://finance. yahoo.com, https://www.ari va.de
Country-specific regulatory requirements	Different taxation, fees, minimum investment amount	Fee from 0 till 2.5% of AUM
Numbers of risk classes	Risk classes: low, medium and high risk levels	There are from 3 till 23 risk classes
Type of required portfolio rebalancing	Rebalance it if a certain threshold value is exceeded	As a rule, threshold or trigger value changes between 3% and 20%

Finally, the investment decisions can be attributed to a number of different factors [13]. The essence of the concept is that the financial instruments of the investment portfolio should be diversified in different terms, types and modifications issued by corporations in different industries and geographical locations [14].

The optimization model and the optimization formula lists include quantitative criteria and performance measures to develop portfolio design and evaluate financial benefits correspondingly (Table 3). The larger Sharp ratio, VaR ratio, Sortino ratio the more successful is an investment strategy. $\bar{r}$ is a portfolio return, r_f is a risk-free interest rate, r_z is an affordable risk level, σ_p is a standard deviation of portfolio, VaR_{CL} means VaR of a certain confidence level. The differences between the RAs relate the specific company, they are driven by the business model and focus on a specific investor's needs.

'Algorithmic advice refers to the automation of professional advice giving by expert systems interacting with consumers instead of highly trained specialists', because the investors who believe that AI is more capable than human intelligence are more likely to adopt algorithmic advice of RA [15]. 'Algorithm aversion is higher when human uniqueness is of great relevance and the tasks are intuitive, subjective but they are not quantifiable and objective' and 'consumers believing that AI is more capable than human intelligence will only be more likely to adopt algorithmic advice when perceived task complexity is high' [15]. The studies on AI advice suggest that getting maximum accuracy in decision-making is a main incentive for investors to use RA. FinTech managers

Table 3. Performance measures for investment portfolio.

Performance measures	Formula
Simple annual return $\bar{r}$	$\bar{r} = \frac{\sum_{i=1}^{n} r_i}{n}$
Annual volatility σ_p	$\sigma_p = \frac{\sum_{i=1}^{n}(r_i - \bar{r})^2}{n}$
Annual VaR $VaR_{annual}^{95\%}$	$VaR_{annual}^{95\%} = (5\% - quantile(r_i) + 1)^{12} - 1$
Maximum drawdown MD	$MD = -minimum\left(vector\left(\frac{\Pi_i(1+r_i)}{cummul\,max\,\Pi_i(1+r_i)} - 1\right)\right)$
Sharp ratio SR_p	$SR_p = \frac{r_p - r_f}{\sigma_p}$
VaR ratio VR_p	$VR_p = \frac{r_p - r_f}{VaR_{CL}}$
Sortino ratio SO_p	$SO_p = \frac{r_p - r_z}{\sqrt{\frac{1}{n}\sum_{t=1}^{n}[\max(r_z - r_t; 0)]^2}}$

may divide the customers into segments according to different beliefs about AI and offer different types of investment and insurance advice to different segments of customers.

'The majority of consumers still express a preference for human financial advisors, because of RAs' lack of a "human touch" and a human ability to understand and personalize investment advice to the consumers unique financial situation' [16]. Thus, conversational as opposed to non-conversational RAs increase the probability for consumers to follow portfolio recommendations even if the advice is inconsistent with actual risk profile or includes larger annual fees. Conversational RAs (i.e., possessing dynamic, dialogue based, and turn-taking communication features) and nonconversational RAs (i.e., possessing static, self-report, and one-way communication features) have distinguishing features. Therefore, RA anthropomorphic design can influence positively on investor's service satisfaction and RA's performance; higher levels of affective trust in a conversational compared to a non-conversational RA increases investors' willingness to accept a recommended financial portfolio [16].

Decision support tool depends greatly on its usability and unwillingness to engage the manage investment questions with RA [17]. Requirements for RA's design principles consist of (1) ease of use (to ensure ease of interaction with the RA); (2) work efficiency (to support users' ability to achieve their goals in expected time); (3) information processing and cognitive overload (to assist users with information processing using video with simplified language of explanation); (4) advisory transparency (to provide disclosure of costs and assets). It is necessary to use different RAs designs to improve the services and user experience.

Financial mistakes made by inexperienced investors, 'who are the largest part of the population, vary for different individuals (e.g., a person has low statistics skills) and in different situations (e.g., stress, cognitive overload)', therefore, inexperienced investors need a good decision support [18]. Multi-modal monitoring includes measurement of users' physiological states (e.g., arousal or cognitive load). It helps to assess the user internal states, incentives and risk attitude.

It means to examine how develop the optimal design of a robo-advisor for guaranteed income estimator for the investor using risk premium under acceptable financial risk. RA can provide such advice to the investors about different portfolios with less or no risk.

Discretion is an investors' ability to override robo-advisors' recommendation, 'robo-advisors that allow for more discretion let investors modify the portfolio weights the algorithm proposes' [19]. Investors can also choose whether proposed algorithm recommendations should be implemented or changed manually. However, to the best of our knowledge, a robo-advisor has no built-in insurance premium module to create incentives for investors to start investing.

3 Models of Robo-Advisor for Different Goals of Investor

The client's risk profile and investment goals are determined before the investment process. In other words, it is what goal the person wants to achieve by the means of the investments in the time horizon. The investor should answer some questions. The answers form the basis for his/her psychological and investment characteristics, it highlights the risk propensity. He/she is also asked to determine the particular investments goals (e.g. to buy a new house, to save money for children's education), because "people have different mental accounting for each investment goal, and the investor is willing to take different levels of risk for each goal" [20]. However, it is suggested to choose not only the primary goal, but also several additional ones. Investors' characteristics can be used to determine the clusters of investors [21]. Diversification, i.e. the using of various investment instruments for different sectors of the economy, occurs not only for one investment portfolio but also for several investment portfolios based on the person's specific goals.

According to the goal, an investor may have a different attitude to risk:

- risk aversion with risk minimization (investors not inclined to take risks of retirement savings);
- risk seeking with return maximizing (investors inclined to take risks launching a startup);
- risk neutral with a desire to achieve minimum risk with maximum return (investors neutral to take risks of savings for a new home).

In addition to profitability, an investor also has to consider the risk associated with the portfolio of financial instruments. According to the Markowitz model, the risk is expressed as the standard deviation σ_p of each financial instrument. The σ_p value is the level of acceptable portfolio risk for the investor. Also, considering the standard deviation of financial instruments, it is necessary to analyse the correlation between profitability of different financial instruments r_{ij}. As a result, we can present the risk of the entire portfolio by the formula (1), where X – matrix of finance instruments' shares, X' – transposed matrix, V^2 – matrix of variations of financial instruments, V_{ij} – matrix of covariations:

$$\sqrt{X^2V^2} + X'V_{ij}X = \sigma_p \qquad (1)$$

The mathematical model for optimal portfolio of financial instruments for an aggressive investor with maximum efficiency $M_p = R'X$, in which the portfolio risk does not exceed a given value σ_p, and considered all restrictions on the portfolio, has the following form (R' - transposed matrix of profitabilities of financial instruments):

$$\begin{cases} M_p \to max; \\ \sigma_p = const; \\ \sum_{i=1}^{n} x_i = 1; \\ x_i > 0, i = 1, \ldots, n. \end{cases} \tag{2}$$

The inverse problem in portfolio optimisation relates to the choice of the portfolio structure with higher or equal expected return M_p with minimal risk σ_p. Consequently, we create a portfolio for a conservative investor. In this case, the mathematical model for the problem has the form:

$$\begin{cases} \sigma_p \to min; \\ M_p = const; \\ \sum_{i=1}^{n} x_i = 1; \\ x_i > 0, i = 1, \ldots, n. \end{cases} \tag{3}$$

Developing a portfolio for a risk-neutral investor, risk minimisation and profit maximisation are simultaneously occured. Thus, we receive the following mathematical model for the problem (4):

$$\begin{cases} \frac{\sqrt{X^2 V^2} + X' V_{ij} X}{R'X} \to min; \\ \sum_{i=1}^{n} x_i = 1; \\ x_i > 0, i = 1, \ldots, n. \end{cases} \tag{4}$$

Next, we consider the architecture of a Robo-advisor based on open data about currency of crypto assets (Fig. 5) for drawing up investment plans [22–25].

In the research, we can choose the cryptocurrency funds since they have low correlation to traditional assets, such as gold or stocks. The cryptocurrency funds can be used for portfolio diversification and creating the independent investment portfolio for a risk-averse investor using advantages investing in commodities and currencies in the financial market [26].

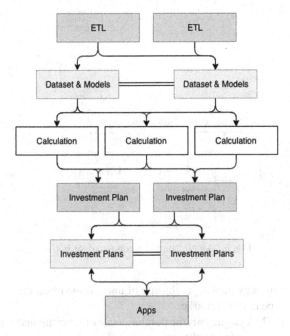

Fig. 5. Architecture of a robo-advisor.

4   Strategies of Robo-Advisor Bot for Different Financial Instruments

Completing an experiment, a software robo-advisor bot was developed using Python Anaconda and Python Jupyter technology. In its final form our robo-advisor (RA) has the architecture shown in Fig. 6 [20].

The customer has to register or, if he/she is already registered, enter the service on the website or in application. The advantage of the developed application is a short registration procedure without gathering confidential information. All you need if you want to register is to confirm your mobile phone number, and then you are given 3 days to use and test the robo-advisor at demo account. Next, the user has an opportunity to choose one of the investment plans based on personal preferences. All investment portfolios are divided into 3 categories: low, medium and high-risk attitude. After choosing one of them, several investment plans become available to customer. In this application customer can find various services, ranging from creating a diversified investment portfolio to purchasing the services of robo-advisor, which trade automatically on platforms (e.g. Binance) in real time. Using any product, client can see in real time all the operations performed by the bot, open and closed transactions, the time and date of transaction closing, purchase and sale of any financial instrument.

Each investor can choose different investment strategies. The most effective of them are:

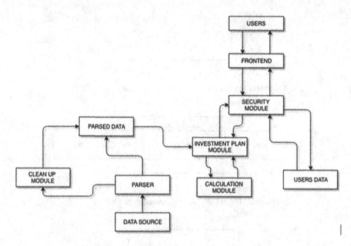

Fig. 6. Architecture of a robo-advisor bot.

1) Buy-and-Hold strategy means purchasing of undervalued financial instruments and their sale at the peak of the market.
2) Moving average (MA) strategy is the purchase/sale of financial instruments when the trend changes, as indicated by the intersection of moving averages for the short-run, long-run, and medium-run periods.
3) Relative strength index (RSI) strategy determines the signal to buy/sell, when the indicator has touched the level 30% (70%) or it is in the oversold/overbought zone, the MACD indicator is above/below the zero level.
4) Support and resistance strategy determines the extremes of the financial instrument price, defining the entry/exit points on the market.

Let us consider the performance of the developed robo-advisor bot. To begin with, it is necessary to determine the closing days and limits of purchase/sale of financial instruments for various strategies (Fig. 7). Thus, when the price of a financial instrument falls to 70% of the initial value, a sale occurs. When the price increases by 30%, compared to the initial value, an automatic sale of the financial instrument happens.

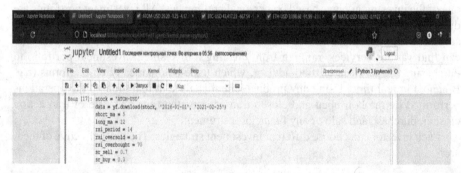

Fig. 7. Algorithmic trading limits for buying and selling financial instruments (code snippet).

After that, data about the closing price of the financial instrument in a specified period are collected and the necessary parameters for calculating the rate of return and risk of this asset are determined using metrics of return and risk from previous Sect. 3 (Fig. 8).

```
data['MA' + str(short_ma)] = data['Close'].rolling(short_ma).mean()
data['MA' + str(long_ma)] = data['Close'].rolling(long_ma).mean()
data['return'] = data['Close'].pct_change()
data['Up'] = np.maximum(data['Close'].diff(),0)
data['Down'] = np.maximum(-data['Close'].diff(),0)
data['RS'] = data['Up'].rolling(rsi_period).mean()/data['Down'].rolling(rsi_period).mean()
data['RSI'] = 100 - 100/(1 + data['RS'])
data['S&R'] = (data['Close']/(10**np.floor(np.log10(data['Close']))))%1
```

Fig. 8. Formulas for calculating rate of return and risk of financial instruments (code snippet).

Next step is preparation of signals' description for buying/selling financial instruments and setting trading frameworks (Fig. 9). These signals are the triggers for algorithm to buy/sell financial instruments when prices decrease/increase by 70%/30% compared to the initial value.

```
BnH_return = np.array(data['return'][start+1:])
MACD_return = np.array(data['return'][start+1:]) * np.array(data['MACD_signal'][start:-1])
RSI_return = np.array(data['return'][start+1:]) * np.array(data['RSI_signal'][start:-1])
SR_return = np.array(data['return'][start+1:]) * np.array(data['S&R_signal'][start:-1])
```

Fig. 9. Description of buy/sell signals for financial instruments (code snippet).

The next step in the chat bot is the calculation of the investor's income from the buying/selling financial instrument using various strategies during definite period (Fig. 10):

```
BnH = np.prod(1+BnH_return)**(252/len(BnH_return))
MACD = np.prod(1+MACD_return)**(252/len(MACD_return))
RSI = np.prod(1+RSI_return)**(252/len(RSI_return))
SR = np.prod(1+SR_return)**(252/len(SR_return))
```

Fig. 10. Calculation asset returns for different investment strategies.

Next, we calculate the risks for various strategies, based on the closing price of financial instruments of previous periods using formula of standard deviation of the samples (Fig. 11).

```
BnH_risk = np.std(BnH_return) * (252)**(1/2)
MACD_risk = np.std(MACD_return)* (252)**(1/2)
RSI_risk = np.std(RSI_return) * (252)**(1/2)
SR_risk = np.std(SR_return) * (252)**(1/2)
```

Fig. 11. Calculation of the risks level of the asset under various investment strategies.

Based on all the obtained results, we made forecasts using our robo-advisor bot for the following period (month) for specified financial instruments using different strategies (Fig. 12).

```
print('доходность и риск стратегии Buy-and-Hold ' + str(round(BnH*100,2))+'% и ' + str(round(BnH_risk*100,2)) + '%')
print('доходность и риск стратегии скользящих средних ' + str(round(MACD*100,2))+'% и ' + str(round(MACD_risk*100,2)) +
print('доходность и риск стратегии RSI ' + str(round(RSI*100,2))+'% и ' + str(round(RSI_risk*100,2)) + '%')
print('доходность и риск стратегии поддержка и сопротивление ' + str(round(SR*100,2))+'% и ' + str(round(SR_risk*100,2)
```

```
<                                                                                      >
доходность и риск стратегии Buy-and-Hold 133.91% и 81.97%
доходность и риск стратегии скользящих средних 156.57% и 81.92%
доходность и риск стратегии RSI 63.66% и 49.85%                    Активация Windows
доходность и риск стратегии поддержка и сопротивление 85.59% и 67.73%    Чтобы активировать Wind
```

Fig. 12. Comparison of profitability and risks for different investment strategies of the specified financial instrument

Our developed software: (https://drive.google.com/drive/folders/1ks5duDulG3H suNONzm-nCmhpjqKcWkZi?usp=sharing) allows automatically computing the profitability and risk of defined financial instruments using different investment strategies choosen for the investor goals.

Let us consider the return and risk of investing in precious metals: gold stock = 'GC = F', silver stock = 'SI = F', nickel ^N225 by the means of the developed robo-advisor bot (Table 4).

Table 4. Indicators of profitability and risk of investing in precious metals.

Financial instrument	Strategy	Return/risk
Gold	Buy-and-Hold	110.21%/14.82%
	Moving average	94.37%/14.83%
	RSI	96.84%/7.51%
	Support and resistance	106.76%/11.25%
Silver	Buy-and-Hold	114.46%/28.82%

(continued)

Table 4. (*continued*)

Financial instrument	Strategy	Return/risk
	Moving average	95.41%/28.84%
	RSI	94.44%/18.6%
	Support and resistance	123.81%/19.65%
Nickel	Buy-and-Hold	111.83%/20.1%
	Moving average	96.92%/20.12%
	RSI	96.92%/10.73%
	Support and resistance	100.24%/16.06%

In the next stage, we calculated the profitability and risk of investing in oil stock = 'CL = F' under different strategies (Table 5).

Table 5. Indicators of profitability and risk of investing in oil.

Financial instrument	Strategy	Return/risk
Oil	Buy-and-Hold	114.22%/155.92%
	Moving average	202.04%/155.75%
	RSI	nan*%/63.9%
	Support and resistance	146.55%/141.66%

* 'nan' means not announced.

For cryptocurrencies, we calculated the profitability and risk of investing in Ethereum (ETH-USD) and Cardano (ADA-USD) (Table 6).

Table 6. Indicators of profitability and risk of investing in cryptocurrencies.

Financial instrument	Strategy	Return/risk
Ethereum (ETH-USD)	Buy-and-Hold	133.91%/81.97%
	Moving average	156.57%/81.92%
	RSI	63.66%/49.85%
	Support and resistance	85.59%/67.73%
Cardano (ADA-USD)	Buy-and-Hold	216.57%/133.3%
	Moving average	243.89%/133.24%
	RSI	nan*%/97.49%

(*continued*)

Table 6. (*continued*)

Financial instrument	Strategy	Return/risk
	Support and resistance	123.63%/111.36%

* 'nan' means not announced.

For stocks, let's compute the return and risk of investing in Netflix, Inc. (NFLX), Tesla, Inc. (TSLA) (Table 7).

Table 7. Indicators of profitability and risk of investing in shares.

Financial instrument	Strategy	Return/risk
Netflix, Inc. (NFLX)	Buy-and-Hold	139.82%/40.63%
	Moving average	74.62%/40.7%
	RSI	104.02%/19.34%
	Support and resistance	98.53%/32.18%
Tesla, Inc. (TSLA)	Buy-and-Hold	177.19%/58.0%
	Moving average	125.42%/58.14%
	RSI	61.72%/39.97%
	Support and resistance	102.47%/45.72%

On the basis of various and diversified financial instruments, we consider the most attractive taking into account the "risk-return" criterion (Figs. 13, 14, 15 and 16). The lower this indicator is, the more investment-attractive the financial instrument is, since 1% of the return represents a lower investment risk. Risk-return criterion has the following form:

$$RRc = \frac{risk}{return} \tag{5}$$

For risk-averse investors all strategies according to the "risk-return" criterion for precious metals demonstrate the greatest efficiency. RSI (relative strength index), Buy-and-Hold strategies are also effective for Netflix shares. For risk-seeking investors all strategies include oil and cryptocurrencies. For risk-neutral investors Tesla stock is the most appropriate during addressed period. This list of financial instruments is obviously not exhaustive and the preferences of investors can include any other kinds of assets.

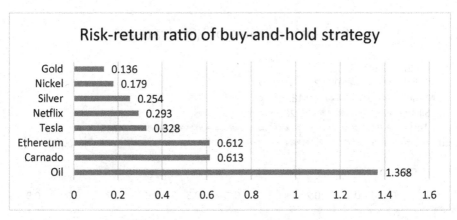

Fig. 13. Risk-return ratio of buy-and hold strategy

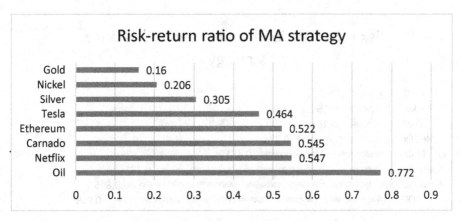

Fig. 14. Risk-return ratio of moving average strategy

In the future, we plan to expand the functions of robo-advisor chat-bot. We are trying to develop of a full-fledged mobile application and website, to expand the range of strategies for providing more convenient and reliable consulting.

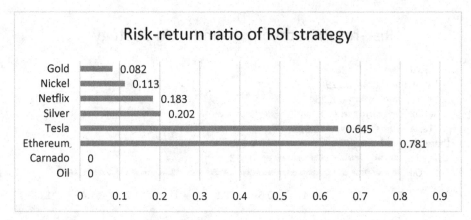

Fig. 15. Risk-return ratio of RSI strategy

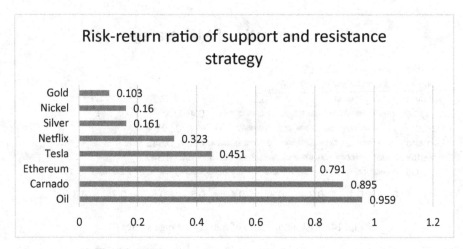

Fig. 16. Risk-return ratio of support and resistance strategy

5 Discussion

Developed software robo-advisor bot is prepared for different risk attitude investors who plan to meet their financial goals. Combining of different financial instruments in one investment portfolio can create unique combination of return and risk ratio, which is not coincide with such metrics of each asset alone. It creates opportunity for individuals to form such structure of their portfolio that is more correspondent to risk-return attitude of persons and their financial goals. The more financial assets of different types/from different industries choose an investor, the more diversified investment portfolio is. It reduces a non-systematic risk of the investors' portfolio. If investor get signal from bot that return of portfolio significant decrease/increase he/she can decide whether to buy/sell the asset(s) or not. It means that our bot is a type of self-service robo-advisor, because it provides information to an investor thereby assisting in the decision-making process.

The bot has also some limitations. Our analysis of risk-return ratio of investment portfolio is based on 252 previous days. It gives information to investor about some average values of return and risk of financial instruments. However, due to some shock events or unpredicted situations the risk-return criteria of investment portfolio can significantly changes. Consequences of such events will be the bias and larger confidence intervals of predicted return of investors. Risk attitude of investors and their goals can change over time and it will generate needs to rebalance portfolio structure by themselves. Our bot does not calculate margins of portfolio rebalancing which can decrease profitability of investment portfolio. It can be directions of our future research in this field.

6 Conclusions

The main types of robo-advisors are defined, including full service RAs (robo-managers that do not require any investor intervention and are fully automated), half-service RAs (a traditional or hybrid service that awaits confirmation of an investment transaction), self-service RAs providing investors freedom to cancel the robo-advisor recommendations. Among the existing RAs, the semi-service RA allows to take into account the investor personalized goals, because the investors can choose whether to implement the recommendations proposed by the algorithm or change them manually.

We developed the software module of an robo-advisor bot using Python Anaconda and Python Jupyter technology. Our developed software allows to calculate automatically the profitability and risk of financial instruments under investment strategies chosen for the investor goals.

The application of our software module reveals that some assets need more complex approach and assessment, taking into account several strategies simultaneously. It will help to make more quality and reliable forecasts. Each investor can choose different investment strategies, among which the most effective are "Buy-and-Hold", the strategy of the moving average, the strategy of the relative strength index RSI and the strategy of support and resistance.

For risk-averse investors all strategies according to the "risk-return" criterion for precious metals demonstrate the greatest efficiency. RSI (relative strength index), Buy-and-Hold strategies are also effective for Netflix shares. For risk-seeking investors all strategies include oil and cryptocurrencies. For risk-neutral investors Tesla stock is the most appropriate. This list of financial instruments is obviously not exhaustive and the investors preferences can include any other kinds of assets.

In the future, we plan to expand the functions of robo-advisor chat-bot. We are trying to develop of a full-fledged mobile application and website, to expand the range of strategies for providing more convenient and reliable consulting.

References

1. Markowitz, H.M.: Portfolio selection. J. Financ. **7**, 77–91 (1952). https://doi.org/10.1111/j.1540-6261.1952.tb01525.x

2. Frydman, C., Camerer, C.F.: The psychology and neuroscience of financial decision making. Trends Cogn. Sci. **20**(9), 661–675 (2016). https://doi.org/10.1016/j.tics.2016.07.003
3. Kahneman, D., Tversky, A.: Prospect theory: an analysis of decision under risk. Econometrica: J. Econom. Soc. **47**(2), 263–292 (1979). https://doi.org/10.2307/1914185
4. Campbell, J.Y.: Restoring rational choice: the challenge of consumer financial regulation. Am. Econ. Rev. **106**, 1–30 (2016). https://doi.org/10.1257/aer.p20161127
5. De Bondt, W.: A portrait of the individual investor. Eur. Econ. Rev. **42**, 831–844 (1998). https://doi.org/10.1016/S0014-2921(98)00009-9
6. AQR Capital Management, Words from the Wise: Harry Markowitz. https://images.aqr.com/-/media/AQR/Documents/Insights/Interviews/Words-From-the-Wise-Harry-Markowitz-on-Portfolio-Theory-and-Practice.pdf. Accessed 12 Jan 2021
7. Bikas, E., Jurevičienė, D., Dubinskas, P., Novickytė, L.: Behavioral finance: the emergence and development trends. Procedia. Soc. Behav. Sci. **82**, 870–876 (2013). https://doi.org/10.1016/j.sbspro.2013.06.363
8. Tobin, J.: The Theory of Portfolio Selection. The Theory of Interest Rates, MacMillan, London (1965)
9. Klingenberger, A., Svoboda, L., Frère, M.: Business model of sustainable robo-advisors: empirical insights for practical implementation. Sustainability (Switzerland) **13**(23), 13009 (2021). https://doi.org/10.3390/su132313009
10. Brunen, A.-C., Laubach, O.: Do sustainable consumers prefer socially responsible investments? A study among the users of robo advisors. J. Bank. Financ. **136**, 106314 (2021). https://doi.org/10.1016/j.jbankfin.2021.106314
11. Duygun, M., Hashem, S.Q., Tanda, A.: Editorial: financial intermediation versus disintermediation: opportunities and challenges in the FinTech era. Front. Artif. Intel. **3**, 629105 (2021). https://doi.org/10.3389/frai.2020.629105
12. Helms, N., Hölscher, R., Nelde, M.: Automated investment management: comparing the design and performance of international robo-managers. Eur. Financ. Manag. **28**, 1028–1078 (2021). https://doi.org/10.1111/eufm.12333
13. Snihovyi, O., Ivanov, O., Kobets, V.: Implementation of robo-advisors using neural networks for different risk attitude investment decisions. In: Proceedings of the 9th International Conference on Intelligent Systems, IS 2018, vol. 8710559, pp. 332–336. IEEE, Funchal (2018). https://doi.org/10.1109/IS.2018.8710559
14. Snihovyi, O., Kobets, V., Ivanov, O.: Implementation of robo-advisor services for different risk attitude investment decisions using machine learning techniques. In: Ermolayev, V., Suárez-Figueroa, M.C., Yakovyna, V., Mayr, H.C., Nikitchenko, M., Spivakovsky, A. (eds.) ICTERI 2018. CCIS, vol. 1007, pp. 298–321. Springer, Cham (2019). https://doi.org/10.1007/978-3-030-13929-2_15
15. von Walter, B., Kremmel, D., Jäger, B.: The impact of lay beliefs about AI on adoption of algorithmic advice. Mark. Lett. **33**(1), 143–155 (2021). https://doi.org/10.1007/s11002-021-09589-1
16. Hildebrand, C., Bergner, A.: Conversational robo advisors as surrogates of trust: onboarding experience, firm perception, and consumer financial decision making. J. Acad. Mark. Sci. **49**(4), 659–676 (2020). https://doi.org/10.1007/s11747-020-00753-z
17. Jung, D., Dorner, V., Weinhardt, C., Pusmaz, H.: Designing a robo-advisor for risk-averse, low-budget consumers. Electron. Mark. **28**(3), 367–380 (2017). https://doi.org/10.1007/s12525-017-0279-9
18. Glaser, F., Iliewa, Z., Jung, D., Weber, M.: Towards designing robo-advisors for unexperienced investors with experience sampling of time-series data. In: Davis, F.D., Riedl, R., vom Brocke, J., Léger, P.-M., Randolph, A.B. (eds.) Information Systems and Neuroscience. LNISO, vol. 29, pp. 133–138. Springer, Cham (2019). https://doi.org/10.1007/978-3-030-01087-4_16

19. D'Acunto, F., Rossi, A.G.: Robo-advising. In: Rau, R., Wardrop, R., Zingales, L. (eds.) The Palgrave Handbook of Technological Finance, pp. 725–749. Springer, Cham (2021). https://doi.org/10.1007/978-3-030-65117-6_26
20. Ivanov, O., Snihovyi, O., Kobets, V.: Implementation of robo-advisors tools for different risk attitude investment decisions. In: CEUR-WS, vol. 2104, pp. 195–206 (2018). http://ceur-ws.org/Vol-2104/paper_161.pdf
21. Kilinich, D., Kobets, V.: Support of investors' decision making in economic experiments using software tools. In: CEUR-WS, vol. 2393, pp. 277–288 (2019). http://ceur-ws.org/Vol-2393/paper_273.pdf
22. Kobets, V., Savchenko, S.: Using telegram bots for personalized financial advice for staff of manufacturing engineering enterprises. In: Ivanov, V., Trojanowska, J., Pavlenko, I., Rauch, E., Peraković, D. (eds.) Advances in Design, Simulation and Manufacturing V. DSMIE 2022. Lecture Notes in Mechanical Engineering, pp. 561–571. Springer, Cham (2022). https://doi.org/10.1007/978-3-031-06025-0_55
23. Kobets, V.M., Yatsenko, V.O., Mazur, A., Zubrii, M.I.: Data analysis of personalized investment decision making using robo-advisers. Sci. innov. **16**(2), 80–93 (2020). https://doi.org/10.15407/scine16.02.080
24. Kobets, V., Yatsenko, V., Popovych, I.: Automated forming of insurance premium for different risk attitude investment portfolio using robo-advisor. In: Ignatenko, O., et al. (eds.) ICTERI 2021 Workshops. Communications in Computer and Information Science, vol. 1635, pp. 3–22. Springer, Cham (2022). https://doi.org/10.1007/978-3-031-14841-5_1
25. Savchenko, S., Kobets, V.: Development of robo-advisor system for personalized investment and insurance portfolio generation. In: Ignatenko, O., et al. (eds.) ICTERI 2021 Workshops. Communications in Computer and Information Science, vol. 1635, pp. 213–228. Springer, Cham (2022). https://doi.org/10.1007/978-3-031-14841-5_14
26. Snihovyi, O., Ivanov, O., Kobets, V.: Cryptocurrencies prices forecasting with anaconda tool using machine learning techniques. In: CEUR-WS, vol. 2105, pp. 453–456 (2018). http://ceur-ws.org/Vol-2105/10000453.pdf

SafeCross - VCA Based Road Traffic Regulations Compliance Detection

Rafał Szklarczyk[1]($\boxtimes$)(ID), Joanna Nikodem[1](ID), Paweł Fałat[1](ID), Konrad Foks[3],
Marcin Caputa[2], and Anna Szklarczyk[2]

[1] University of Bielsko-Biała, ul. Willowa 2, 43-309 Bielsko-Biała, Poland
kinf@ath.bielsko.pl
[2] Elmontaż Sp. z o.o., ul.Ks.Pr.St.Słonki 54, 34-300 Żywiec, Poland
[3] Mechatronika.net Sp. z o.o., ul.Piłsudskeigo 42/1, 43-300 Bielsko-Biała, Poland
https://www.ath.bielsko.pl, https://www.mechatronika.net,
https://www.elmontaz.pl

Abstract. The present paper introduces the concept of video content analysis based system for road traffic regulations compliance monitoring at unguarded railway crossings.

The described system named SafeCross is currently being developed by the consortium of Elmontaż Sp. z o. o. (ltd.) and University of Bielsko-Biała.

Recently it gained the final phase of development. The technical goals have been achieved and some work results are available.

Keywords: SafeCross · Road traffic surveillance · VCA · Road traffic AI analysis

1 Introduction

Railway crossings are specific infrastructure elements, because they are an intersections of different traffic stream kinds.

Accidents in these places usually have dire consequences. Their participants often die as a result of that type of incidents or suffer severe injuries that require long-term and costly treatment. The road vehicles involved in these accidents are usually not repairable and are scrapped, while rail vehicles require costly repairs [5].

There were a lot of dangerous incidents observed recently at the railway crossings and passages in Poland. For example in the year of 2019 205 accidents were registered [1]. In last 6 years before that time the annual amount of registered accidents that occurred on railway crossings and passages varied between 208 and 216 [4].

The greatest number of dangerous accidents occur on category D crossings, which are approximately 60.3% of all rail crossings in Poland [2,3].

Category D crossings are not equipped with traffic safety systems and devices at all, there are only traffic signs present.

J. Maślankowski et al. (Eds.): PLAIS EuroSymposium 2022, LNBIP 465, pp. 36–48, 2022.
https://doi.org/10.1007/978-3-031-23012-7_3

The Polish data is presented as the project is supported by a financial grant from the Polish National Centre for Research and Development.

However the described issues consider not only Poland, but also other regions. In the EU there are about 120 000 railway crossings and about half of them have similar conditions and equipment like Polish category D crossings [3].

The aim of the paper is not to analyze the detailed situation in the whole Europe or even farther but to present the concept and so far results in short. The Polish data are used as an example to show the business and social environment of the project and to prove the relevance of the problem [18] that is to be addressed by the project.

It is assumed that the influence of the project implementation and application will improve the safety. However it would be verifiable after the application of the system in some scale. That would be the time and the place to analyze the situation in more detailed manner and to show the project influence.

The accidents inflict human casualties. A human life is priceless. The accident victims that survive often suffer from injuries. The recovery process takes a lot of effort, time and resources. Sometimes the health damages are irreversible.

Besides the effects on people the accidents cause large material loses too. The cost of each accident varies in different amounts, including, inter alia, elements such as: accident victims' costs, infrastructure administrator costs and costs of the rail company. The infrastructure administrator must cover the costs of repairing the crossing (e.g. damaged road surface, pavement, damaged barriers) and suspend the traffic on a given section of the railway line (no income from making the route available to railway undertakings, costs of introducing alternative routes to the railway users). From the perspective of the railway company, costs are also generated in the form of expensive post-accident repairs, the need to use replacement rolling stock, launching replacement transport, additional work time for train personnel or even loss of customer confidence and income.

The unit cost of the fatal victim is approximately PLN 2.4 million; the unit cost of a seriously injured victim is approximately PLN 3.3 million; the unit cost of a road accident is approximately PLN 1.4 million [6].

All the circumstances are the matter of concern of public opinion, authorities, railway companies, infrastructure administrators and finally maintenance companies and infrastructure contractors.

The Polish Office of Railway Transport (Urząd Transportu Kolejowego) announced the need of situation improvement.

In response of that need the company Elmontaż Sp. z o. o. (ltd.) proposed that implementation of a VCA based system, that would monitor a road traffic regulations compliance at railway crossings, could be the remedy.

The system could additionally help to increase drivers' focus as they approach the crossing.

The above resulted in the initiative of Elmontaż Sp z o. o. and University of Bielsko-Biała. Both institutions as a consortium applied for the grant from the Polish National Centre for Research and Development with the project "System of monitoring and analysis of events allowing to increase safety at level D railway crossings, as well as to identify compliance with applicable road traffic regulations" [3].

When grant was issued, this started the development of the system described in the present paper named SafeCross [11].

2   Assumptions, Requirements and the Concept

The construction of the system must take into account the conditions dictated by the realities of the available equipment, available material and human resources, as well as available knowledge, assuming that the project will result in extending the scope of available knowledge.

The system shall provide an appropriate scalability and reasonable running costs.

According to that the original approach was proposed.

The system topology concept is similar to a cluster-based WSN. See Fig. 1 [12]. However there are also some differences (in the structure and logic as well).

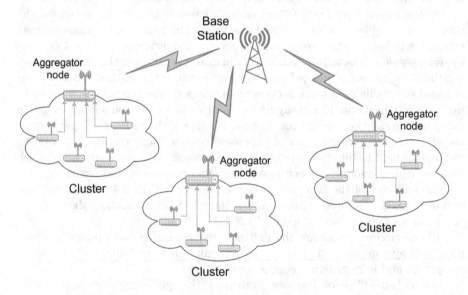

Fig. 1. A cluster-based data aggregation process in the WSNs

In the SafeCross system, CCTV cameras are used in the role of sensors. The signals from the sensors are gathered, preprocessed and relayed by so called Data Acquisition Centers (DACs).

In the role of base station there is the Storage and Processing Center, which has its own cloud computing based architecture.

The processing is divided between DACs and SPC, that the preprocessing performed in DACs optimizes the usage of available hardware and overall computing power.

3   Artificial Intelligence Tasks

The road traffic monitoring is a popular task for artificial intelligence.

A wide survey was presented in [15]. There is a variety of methods used to detect vehicles that use many different types of sensors including Magnetic sensors, Acoustic sensors, Lidars, RF Transceivers, Radars or Wi-Fi Transceivers. The obtained accuracy of detection varies in the range of between about 80% to almost 100% [15].

The detected vehicles could be than classified as different types (ex. cars, vans, trucks etc.).

Some implemented systems offer detected vehicle traffic analysis that could provide real-time output (e.g. alerts, electronic road signs, ramp meters, etc.) as well as extract statistical information (traffic loads, lane changes, average velocity etc.) so it could be used for traffic optimization and equipment control as well (ex. traffic jams avoidance, traffic lights control optimization) [16,17].

However it is hard to find the system that would work in the similar logic as the one needed for railway crossing category D traffic analysis.

SafeCross will introduce the custom AI system that needs to solve the following issues:

- object detection,
- object classification (vehicle detection),
- object behavior classification (road traffic regulation compliance).

4   System Architecture

The discussed system consists of two main components. The data-collecting part and the management part that also performs the analysis and classification of the obtained information. They are treated as separate system layers (Data Acquisition Layer) (Control and Analyse Layer). These layers were made in "distributed" IoT and Cloud Computing techniques. See Fig. 2.

4.1   Data Acquisition Layer

The data-gathering part consists of many different devices. These are mainly cameras operating at railway crossings and recorders that enable quick registration of events captured by the cameras. An important element are dedicated Data Acquisition Centers (DAC), which perform the operation of identifying events, filtering recorded film tracks and communicating with the system that analyzes them by sending it the necessary information. The DAC is also to be equipped with modules enabling basic analysis of events in order to filter them for further processing.

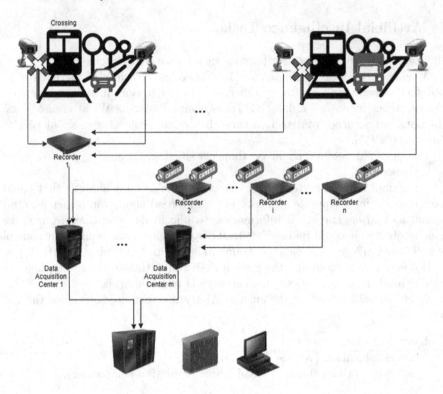

Fig. 2. System architecture

It should also be noted that the system is prepared for modifications in the form of replacement of camera types or installation of other types of sensors, e.g. radar sensors.

This element is therefore part of the issue related to the Internet of Things as a group of distributed sensors and devices with which communication takes place by means of technologies used in the creation of web and mobile applications (e.g. http, REST).

4.2   Control and Analyse Layer

Due to the need to process large amounts of multimedia data, it is made in the cloud technology. This allows for its scaling and gives the opportunity to analyse the delivered materials with appropriate efficiency. At the same time, this system is a place for managing and monitoring the work of the part that collects data. It communicates and configures Data Acquisition Centers (DAC).

4.3   Communication Between Layers

The connection between the layers (collecting and analysing multimedia data) of the system is realized with the use of technology based on the HTTP protocol.

This is used both for downloading data from the DAC and for their configuration. It was assumed that the structures of objects used for communication remained unchanged, which allowed the use of simpler REST services [7] with the OpenAPI standard (Swagger) [8] instead of SOAP or GraphQL services, which were also taken into account when designing. Additionally, mechanisms for identifying and authorizing users to all endpoints have also been implemented.

4.4 Security

Security is a particularly important element in the presented system. Not only at the level of access to it, but also work stability and data consistency. The basis is therefore the implementation of user identification and authorization mechanisms. Internally the system has been isolated. The security level can be farther improved for example by implementing of a mechanism that is based on the use of the JSON Web Token (JWT) standard [9]. Externally the system was secured with the use of proven technologies used in systems communicating via the http protocol which are the standard mechanisms for used programming tools.

System performance has a separate group of security requirements. An additional testing application is planned here, which will enable the control of the DAC centre in terms of efficiency - both at the level of sending ready data as well as for communication with recorders and cameras.

5 Video Content Analysis

The final detection parameters where True Positive Rate (TPR):min. 0.85, True Negative Rate (TNR):min. 0.70 Accuracy (ACC):min. 0.75 by average traffic intensity not greater than 10 vehicles per hour. The above parameters where successfully achieved. However there is possibility to choose a data sample with worse results. Especially when the data was gathered in the conditions of visual signal disturbance (ex. fog). This shows that the system performance relays on the chosen sensors and application environment.

In the initial phase two approaches were taken into consideration in order to detect whether a vehicle has stopped before crossing the trackway or not.

The first approach is motion detection and tracking based and the second is object detection and tracking based.

5.1 Motion Detection and Tracking

In the approach the differential is used to detect moving objects. If the moving object stops, it will fade out and disappear. See Fig. 3.

Fig. 3. Neighbouring frames subtraction in a sequence of movement, stopping and reappearing movement of the car - fading of the vehicle might be observed

The detected moving objects are not classified, but their movement is tracked in modeled path. See Fig. 6.

Fig. 4. Motion detection and tracking (left side - original frame; right side - interpreted frame)

The lack of using machine learning techniques for object classification is causing significant reduction of computing power needs.

5.2   Object Detection and Tracking

The model based on background or neighbouring frame subtraction is however not scale nor random object appearance resistant. Thus, another approach for image recognition is needed to distinguish between road vehicles and other objects, which are not subject of the research. There are plenty of entries in the literature that deal with traffic monitoring systems, a wide survey of which is given in [15]. A smaller number of works apply to traffic flow estimation, which deals not only with road object detection, but also the determination of movement direction [14] and anomaly detection [13]. In this work we apply an efficient and fast deep learning based method to object detection consisting of MobileNet or YOLO architecture and the Single Shot Detector (SSD) framework, as implemented in OpenCV 4.5.1 . Figure 5 and Fig. 6 shows two sample frames with detected desired (car) and random (person) objects together with a confidence parameter printed at the top of a bounding rectangle of each.

Fig. 5. An object detected as a car with 100% confidence

The detected objects are classified and then they are tracked. The Fig. 7 shows the interpreted movie frame.

5.3   Results

The tests of both approaches were made in controlled environment. The tests results are shown in Table 1.

The both approaches gave the satisfying results in controlled environment. In the case of the test number 3 the movement detection approach failed. It turned out that the recorded event occurred in the exact moment of wind squall, which caused a shaking camera effect which disturbed the actual motion detection. The object detection approach performed better as it is resistant on such circumstances.

Fig. 6. Two objects detected: a person (41,35%) and a car (99,98% of confidence)

Fig. 7. Object detection and tracking (left side - original frame; right side - interpreted frame)

Table 1. The results of tests in controlled environment tables.

Test number	Date	Expected result	Motion detection result	Object detection result
1	2021-11-27 11:58:53	OK	OK (TN)	OK (TN)
2	2021-11-27 11:59:41	NOK	NOK (TP)	NOK (TP)
3	2021-11-27 12:05:36	OK	??? (Unknown)	OK (TN)
4	2021-11-27 12:11:07	NOK	NOK (TP)	NOK (TP)
5	2021-11-27 12:15:48	OK	OK (TN)	OK (TN)
6	2021-11-27 12:20:41	NOK	NOK (TP)	NOK (TP)
7	2021-11-27 12:25:01	OK	OK (TN)	OK (TN)
8	2021-11-27 12:30:08	NOK	NOK (TP)	NOK (TP)
9	2021-11-27 12:35:03	OK	OK (TN)	OK (TN)
10	2021-11-27 12:40:09	NOK	NOK (TP)	NOK (TP)
11	2021-11-27 12:45:04	NOK	NOK (TP)	NOK (TP)
12	2021-11-27 12:49:59	OK	OK (TN)	OK (TN)
13	2021-11-27 12:55:27	OK	OK (TN)	OK (TN)

In the real world application, which have been implemented at two real railway crossings, the object detection approach performed significantly better. It is more high traffic intensity level resistant and more other interference tolerant.

The exact comparison of performance of the two approaches is not presented in the paper as it is dependent on real world data samples. It could be dramatically different for a various data samples. The results depend on multiple factors, especially on visual channel disturbance, traffic intensity, geographic topology of installation site, properly installed and configured infrastructure and system configuration.

During the development phase many program versions of both approach were tested and compared. Their results differed. Which shows that not only the method influences the results but also the advance of development. There are also some program parameters that could be adjusted as well.

Generally the object detection approach performed about twice better as the movement detection approach. There were a lot of real world cases when the movement detection approach algorithm could not interpret the data, but it had slightly better TPR and worked over 20 times faster then the object tracking approach (ex. computing time 1.347 s of the first approach vs 46.539 s of the second approach). However the TNR was a lot better for object detection approach.

There were also some cases when the object detection approach failed because the object present on recorded material could not be recognized by the used neural network implementation.

6   Summary

The previous chapters provide a description of the concept and a framework description of the proposed solution architecture as well as the current achievements in the development of the project.

In the term of the Design Science Research [18] the project introduces two kinds of artifacts:

1. the system itself
2. the AI mechanisms

The first one should be considered as a generic architecture concept that could be used to gather information about events and to assess them. It implements a structure with a plugin capabilities, where both event source and assessment mechanisms could be replaced.

The second consists of mechanisms described in Sect. 5.

Both kind of artifacts fulfills 7 guidelines for a design science research [18]:

1. **Design as an artifact** -both kinds of artifacts were successfully implemented and tested.
2. **Problem relevance** - as described in Sect. 1.
3. **Design evaluation** - was performed and the project performance satisfied requirements described in Sect. 5.
4. **Research contributions** - the research results and the design concept are described in the present paper. The project was already demonstrated in business and social environment.
5. **Research rigor** - the requirements for the designed artifacts were defined and the artifacts were tested against them.
6. **Design as a search process** - the final results were obtained by implementing and testing of various approaches and program versions. Finally the two approaches were chosen, implemented and described.
7. **Communication of research** - the present paper describes the project for the technology-oriented audience. The work was also presented for both technology-oriented management-oriented audiences on conferences (ex. [19, 20]). Some other information actions has been performed like website publication and providing of marketing materials.

There are plenty of scenarios of farther research and development of the presented system. An Image stabilization could be implemented to improve the motion detection implementation. A hardware acceleration or neural network simplification can be tested to improve the object identification efficiency. Finally some other approaches could be proposed like one based on a combination of two described in the present paper.

Additionally, experience will be gathered to enable the development of the system in the desired direction. Drivers' behavior or other circumstances affecting safety may be observed. Thanks to this, it will be possible to draw conclusions regarding the applicable procedures and traffic organization.

IT system solutions have the feature that they can be relatively easily modified. Changes to the software are possible. The project has a great potential for expansion and modernization resulting from these features.

References

1. Szymajda M.: W 2019 mniej wypadków na torach, ale więcej ofiar śmiertelnych. Rynek kolejowy (2020). https://www.rynek-kolejowy.pl/wiadomosci/w-2019-mniej-wypadkow-na-torach-ale-wiecej-ofiar-smiertelnych-95197.html
2. Baza Przejazdów Kolejowych - PKP Polskie Linie Kolejowe S. A
3. System monitoringu i analizy zdarzeń pozwalający na podniesienie bezpieczeństwa na przejazdach kolejowych kategorii D, a także identyfikujący przestrzeganie obowiązujących przepisów ruchu drogowego. NCBR Grant Application nr POIR.04.01.04-00-0048/20
4. Urząd Transportu Kolejowego. Bezpieczeństwo w 2019 roku - wstępne podsumowanie. https://utk.gov.pl/pl/aktualnosci/15664,Bezpieczenstwo-w-2019-roku-wstepne-podsumowanie.html
5. Najwyższa Izba Kontroli. Bezpieczeństwo ruchu na przejściach i przejazdach kolejowo-drogowych (2018) https://www.nik.gov.pl/plik/id,12954,vp,15363.pdf
6. Krajowa Rada Bezpieczeństwa Ruchu Drogowego. Wycena kosztów wypadków i kolizji drogowych na sieci dróg w Polsce na koniec roku 2018, z wyodrębnieniem średnich kosztów społeczno-ekonomicznych wypadków na transeuropejskiej sieci transportowej (2019) https://www.krbrd.gov.pl/wp-content/uploads/2020/12/Wycena-kosztow-wypadkow-i-kolizji-drogowych-2018.pdf
7. REST.https://www.service-architecture.com/articles/web-services/representational-state-transfer-rest.html
8. OpenAPI (swagger) specification. https://swagger.io/specification/
9. JWT standard https://jwt.io/introduction
10. NCBR. http://ncbr.gov.pl
11. NCBR. Lista ocenionych projektów złożonych w ramach Programu Operacyjnego Inteligentny Rozwój 2014–2020 działanie 4.1/poddziałanie 4.1.4, Projekty aplikacyjne". https://archiwum.ncbr.gov.pl/fileadmin/POIR/1_4_1_4_2020/Lista_rankingowa_1_4.1.4_2020.pdf
12. Nasridinov, A., Ihm, S.-Y., Park, Y.-H.: Skyline-based aggregator node selection in wireless sensor networks. Int. J. Distrib. Sens. Netw. (2013). https://doi.org/10.1155/2013/356194
13. Sodemann, A., Ross, M.P., Borghetti, B.J.: A review of anomaly detection in automated surveillance. IEEE Trans. Syst. Man Cybern. Part C (Applications and Reviews) **42**(6) (2011). https://doi.org/10.1109/TSMCC.2012.2215319https://doi.org/10.1109/TSMCC.2012.2215319
14. Fedorov, A., Nikolskaia, K., Ivanov, S. et al.: Traffic flow estimation with data from a video surveillance camera. J. Big Data **6**, 73 (2019). https://doi.org/10.1186/s40537-019-0234-z
15. Won, M.: Intelligent traffic monitoring systems for vehicle classification: a survey. IEEE Access **8**, 73340–73358, (2020). https://doi.org/10.1109/ACCESS.2020.2987634https://doi.org/10.1109/ACCESS.2020.2987634
16. Koutsia, A., Semertzidis, T., Dimitropoulos, K., Grammalidis, N., Georgouleas, K.: Intelligent traffic monitoring and surveillance with multiple cameras. (2008). https://doi.org/10.1109/CBMI.2008.4564937https://doi.org/10.1109/CBMI.2008.4564937

17. Biswas, S.P., Roy, P., Patra, N., Mukherjee, A., Dey, N.: Intelligent traffic monitoring system. In: Satapathy, S.C., Raju, K.S., Mandal, J.K., Bhateja, V. (eds.) Proceedings of the Second International Conference on Computer and Communication Technologies. AISC, vol. 380, pp. 535–545. Springer, New Delhi (2016). https://doi.org/10.1007/978-81-322-2523-2_52

18. Hevner, A.R., March, S.T., Park, J., Ram, S.: Design science in information systems research. MIS Q. **28**, 75–106 (2004). http://citeseerx.ist.psu.edu/viewdoc/download?doi=10.1.1.103.1725&rep=rep1&type=pdf

19. Szklarczyk R.; Izydor D.; Presentation at the conference "Bezpieczeństwo na kolei", Jastrzębia Góra (2021) http://bezpieczenstwonakolei.pl

20. Szklarczyk, R.: System monitoringu i analizy zdarzeń pozwalający na podniesienie bezpieczeństwa na przejazdach kolejowych kategorii D, a także identyfikujący przestrzeganie obowiązujących przepisów ruchu drogowego. IX Conference, Rozwiązania Skrzyżowań Kolei z Drogami Kołowymi w Poziomie Szyn w Aspekcie Prawnym, Ekonomicznym i Technicznym". Kroczyce (2022) https://www.przejazdy.eu

Creativity and Innovations

Promoting Creativity with Social Media Knowledge Discussion Groups: Exploring the Moderating Role of Knowledge-Oriented Leadership

Anam Nusrat[1], Yong He[1(✉)], and Adeel Luqman[2]

[1] School of Economics and Management, Southeast University, Nanjing 210096, China
hy@seu.edu.cn
[2] College of Management, Research Institute of Business Analytics and Supply Chain Management, Shenzhen University, Shenzhen 518060, China
adeel@ustc.edu.cn

Abstract. This study aims to examine the impact of social media knowledge discussion groups (SMKDGs) on creativity through knowledge utilization, and the impact of SMKDGs on employee creativity through information elaboration in a multi-level model. In addition, the moderating effect of knowledge-oriented leadership (KOL) was also tested at two levels. Multisource and multi-wave data were collected from 356 employees in 61 workgroups and used for empirical analysis. The results support our hypothesis and confirm the mediating role of knowledge utilization and information elaboration at two levels. In addition, KOL positively moderates the mediating effects of knowledge utilization and information elaboration at both levels. These findings complement research on SMKDGs and creativity and have instructive value for teams to effectively use social media to enhance the team and employee creativity in the future.

Keywords: Enterprise social media · Knowledge utilization · Leadership · Innovative performance · Multi-level · Knowledge management

1 Introduction

Knowledge and creativity are key factors not only for organizational competitiveness, but also for companies' adaptability, resilience, and productivity [1]. In fact, there is a reciprocal relationship between organizational capacity to maintain its ability to generate knowledge, ideas and provide mechanism to enhance employee's knowledge and innovative ability [2]. Knowledge management (KM) is a discipline that "promotes an integrated approach to identifying, capturing, evaluating, retrieving, and sharing all the enterprise's information assets included databases, documents and procedures, among others" [3]. Thus, in a dynamic and competitive market where employee creativity is a crucial factor, the use of social media has inspired many organizations to coordinate and leverage knowledge to promote sustainable competitive advantage over competitors [4].

© The Author(s), under exclusive license to Springer Nature Switzerland AG 2022
J. Maślankowski et al. (Eds.): PLAIS EuroSymposium 2022, LNBIP 465, pp. 51–64, 2022.
https://doi.org/10.1007/978-3-031-23012-7_4

Social media tools are designed for organizational settings to facilitate knowledge sharing among employees and have the potential to foster creativity [5]. It is well noted that social media positively contribute to promotes innovations and [6] creating sustainable advantage for organizations [7, 8]. However, effective technological settings to promotes right mix of KM system to leverage organization wide knowledge, kindle information among the employee for creativity is still challenge for the organizations.

In order to solve organizational problems, it is important to generate relevant knowledge and effectively disseminate it under a unified discussion platform. For example, according to [9], "dispersedness of knowledge is inextricably linked to the problem of designing communication structures." However, existing social media research is limited in providing mechanisms to facilitate effective KM systems that facilitate organizations creativity, thereby revealing certain limitations and research gaps. First, extant research mainly examines social media use in isolation (e.g., [6, 10]). However, we examined the extent to which social media knowledge discussion groups (SMKDGs) in organizations enhance the employee knowledge base and impart creativity at both individual and team levels. We argue that SMKDGs promote two communications that are much more effective and provide particular solutions by aligning the directionality of group discussion. SMKDGs are an ideal solution for establishing the right mix KM system and provide more frequent and direct communications. SMKDGs help in generating more content to provide vast knowledge-based and facilitate knowledge utilization and promote employee creativity. Hence, it is reasonable to investigate the impact of SMKDGs on creativity from a multi-level perspective.

Second, previous studies have mainly considered isolation mechanisms at the individual or team level, such as employee psychological meaning, availability and safety [11], instrumental and expressive relationships [6], and social bonds [7] in promoting employee performance [12]. However, existing academic research has not reached a consensus on whether social media empowers creativity through individual or team-level structure. For example, many tasks in business are done by work teams, and creativity research is moving from the individual level to the team level. Team creativity is not a simple sum of employee creativity, its formation mechanism is more complex. Few studies have tended to explore how social media use affects team creativity. For example, the use of social media facilitates communication within teams to facilitate team knowledge sharing, and possibly team creativity. However, diverse information sources and team member backgrounds have also been identified as important sources of creativity [13]. However, the other side claimed that by providing various inputs or stimuli, the cross-level relationship between the team environment and individual performance was recognized [1]. Therefore, when the mechanisms of creativity at the team and individual levels are not clear, both the team level (i.e., knowledge utilization) and the individual level (i.e., information elaboration) can be examined to promote creativity. We adopted an information exchange perspective [14] and combine creativity at the team and individual levels to examine creativity [15]. We assumed that since SMKDGs is rich sources of information may enhance information exchange among team members and facilitate team knowledge assimilation and increase the accessibility of information, we investigated the mediation of knowledge utilization at team level and information elaboration at individual between SMKDGs and creativity at team and individual respectively.

Third, existing research focuses on boundary conditions such as competitive intelligence [16], social media capabilities [17], and relational and intuitive decision-making [18]. Because SMKDGs can support team knowledge utilization and employee information presentation. However, a team climate that encourages learning and collaboration is also important for predicting team knowledge utilization and employee information elaboration[19]. Although leadership is very important in developing knowledge and creativity, scholars have recently begun to explore the role of leadership in knowledge management. For example, leaders in organizations are called innovators and facilitators in technological environments [20]. Therefore, the authors emphasize identifying leadership types to clarify roles in the KM system. We answer the call for research to explore the role of knowledge-oriented leadership (KOL) in fostering creativity through SMKDG. KOL is when leaders create an atmosphere at the individual and team level that promotes learning and supports the learning process, so that team goals can be accomplished quickly and efficiently [21]. Specifically, we investigated the moderating effect of KOL on the association between SMKDGs and individual and team creativity. We argue that when KOLs consciously interact with SMKDG to create an organizational environment conducive to knowledge creation, their professional knowledge management behaviors can make members perceive that the work environment supports KM activities [21, 22].

Based on the previous discussion, emphasizing the need for a better understanding of the role of SMKDGs in knowledge utilization and knowledge elaboration to promote creativity, this study proposes two research questions (RQs): RQ1: Do knowledge utilization and knowledge elaboration mediate between SMKDGs and creativity at the team and employee levels? RQ2: What is the role of KOLs in contextualizing the effects of SMKDGs in promoting creativity? (see Fig. 1 for details) This research is rooted in a Knowledge-Based View (KBV), which represents a strategic asset of firm competitiveness [23]. According to KBV, an organization is the embodiment of a knowledge-bearing entity that manages knowledge resources, which are considered dynamic capabilities [24].

To address the above research questions, the current study contributes to existing research in several ways. First, previous research has rarely considered the connections between the three isolated domains of social media-based KM systems, leadership, and creativity. Building links between these three fields, we contribute theoretically to the knowledge management and leadership literature, and pave the way for knowledge-intensive industries to foster creativity through SMKDGs and KOLs. Second, previous research has discussed various leadership characteristics, such as servant leadership [25] and authoritarian leadership [26]. However, given the nature of the research context, the current study discusses a unique type of leadership behavior within the KM initiative, known as KOL [27]. Our study thus contributes to the existing leadership literature supporting creativity and knowledge transfer among employees embedded in social networks. Finally, our study provides a new mechanism of knowledge utilization and knowledge elaboration from a multi-level perspective, which helps SMKDG to positively influence the creativity of teams and employees.

2 Theory and Hypotheses

Proposed research model is presented in Fig. 1.

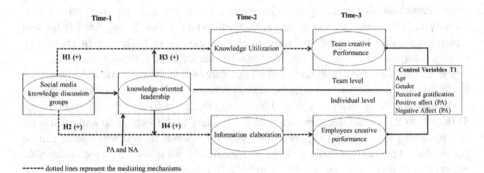

Fig. 1. Proposed research model

SMKDG refers to knowledge management systems that provide rich information through formal and informal discussions [28]. Social media are web-based platforms that support knowledge management activities [6, 11]. Social media can achieve superior work outcomes by helping employees build information-rich knowledge management systems. Social media-based communication provides an instant information-rich solution to management problems through information communication [29]. The increase in the amount of user-generated content on social media platforms may enable employees to make expressive and instrumental connections [12], which further helps employees understand shared content on social media [1]. We believe that SMKDG can provide relevant and rich information to individual team members, which may enhance team knowledge utilization, which is the process by which team members learn knowledge, including knowledge acquisition, knowledge transfer, and new knowledge implementation [30]. As a mechanism for integrating different knowledge sources into ideas, team knowledge utilization is designed to motivate team members to learn actively and help them accomplish work tasks [31]. We infer that SMKDG teams are indirectly associated with team creativity through knowledge utilization. Therefore, we make the following assumptions:

H1: SMKDG has indirect association with team creativity via knowledge utilization

Employee information elaboration refers to the process by which employees integrate their own information with that of others, which is knowledge creation at the individual level [32]. Existing literature points out that information refinement has undergone the development of process, cognitive mechanism, and process-cognitive mechanism perspectives [33]. From the perspective of process cognitive mechanisms, [34] pointed out that the information elaboration process is a complex form of communication in which team members elaborate personal views and discuss their views with each other, deciding how to use and articulate information resources to solve problems under the action of cognitive mechanisms something wrong. Therefore, we argue that SMKDG enhances employees' information elaboration and integrates information relevant to their work tasks, enabling them to generate ideas [35] For example, employee information elaboration is often considered a panacea for driving employee creativity, as the process of information elaboration enables employees to obtain richer information to stimulate their creativity [5].

H2: SMKDG has indirect association with employee creativity via knowledge elaboration

As mentioned earlier, KOLs provide their employees with an atmosphere that promotes learning and supports the learning process at the individual and team level (e.g., [36]). We believe that KOLs will not only focus on self-learning and set a learning example for team members, but also focus on knowledge management, guide team members to participate in knowledge and information management activities, and improve team creativity. SMKDG is a suitable platform for KOLs to create learning goal orientation, create a learning atmosphere, support the learning process at both team and individual levels, and be role models. Since SMDGs allow users to interact with different groups of people with different expertise and backgrounds, KOLs may have to interact to enhance the knowledge learning atmosphere of individuals and teams. With high-level KOLs, team and individual members believe that their efforts will be rewarded, and will be committed to collective goals and actively participate in learning and knowledge activities [17]. Therefore, we believe that effective KOLs may improve KM practices through self-modeling and create an atmosphere that promotes knowledge utilization and information refinement behaviors of employees. Therefore, we assume the following:

H3: A knowledge-oriented leadership positively moderate the indirect relationship between SMKDG and team creativity via knowledge utilization in such a way that the indirect relationship is stronger when knowledge-oriented leadership is higher vs. low

H4: A knowledge-oriented leadership positively moderate the indirect relationship between SMKDG and employee creativity via knowledge elaboration in such a way that the indirect relationship is stronger when knowledge-oriented leadership is higher vs. low

3 Method

3.1 Data Collection and Procedure

Data were collected from project teams from four multinational software and high-tech companies using a multi-source and multi-wave survey design. We chose technology companies because of knowledge-intensive nature and creativity is paramount in their strategies that needs to be discussed before and during projects [37]. We invited 89 teams across all teams working in different departments and embodying formal hierarchies such as leaders and team members. Every small team needs to complete complex tasks together while working in a small team and competing in a competitive marketplace. To avoid common method bias (CMB), data were collected in three waves. In the first wave, all 89 teams targeted, however 71 teams, including 452 members responded with a response rate of 79.7%. During this phase, employees assess knowledge-oriented leaders. Additionally, employees provided data on their perceptions of SMKDG and control variables such as demographic age, gender, perceived usefulness, and personality traits (PA and NA). In the second phase, we invited 71 teams that provided data in the first phase, however, 65 teams responded, including 398 respondents, with a response rate of 91.5%. In this this phase about a month later, employees rated their perceptions of knowledge utilization and elaboration. Finally, in the third phase, we re-invited the team that completed the investigation in phase 2. In the third phase, the sample of available teams was 61 teams with a total of 356 respondents. In the third phase of data collection, leaders assessed their teams' and individual employees' perceptions of creativity. In addition, as a token of appreciation, we provided participants with gifts, such as books or souvenirs, worth about \$5 ($\approx$ 30 yen).

4 Measurements

To ensure the validity and reliability of the study, we used existing measurement scales in our study. However, we modified the project slightly based on the context. All scales were measured using a five-point Likert scale (1 = "strongly disagree", 5 = "strongly agree") unless otherwise stated. SMKDG was assessed using an eight-item scale adapted from [38]. An example project is "I often have heterogeneous discussions with colleagues on a variety of personal topics to address technical issues that enhance my knowledge through the ESM network". Employee information elaboration was captured using the four-item scale suggested by. An example item is "At work, I strive to consider all task-related information to generate the best solution." We use three item scales proposed by [39] to measure the knowledge utilization of the team. An example item is "Team members' task-related expertise and skills are fully utilized in our team's activities." Employee creativity was assessed using a four-item scale developed by [40]. An example item is "[--NAME---] generates ideas that are revolutionary in the field" Finally, we adopted the four-item scale of team creativity suggested by [41]. An example item is "How well does your team generate new ideas?".

4.1 Common Method Bias

While we collected data from multiple sources and multiple waves using a time-delay approach, since our data were self-reported, we further employed various statistical measures to determine the likelihood of CMB in our data. First, as suggested by [42]. Harman's one-way test for common method variance. The results of Harman's single factor test show that the variance explained for the first factor is 26.85%, which is well below the critical value (that is, 50) for the total variance explained for the 71.24% factor, indicating that CMB is not a problem. Our data. Secondly, we checked for the presence of multicollinearity effects with variance inflation factor (VIF) and tolerance. Consistent with previous studies [43], the results showed VIFs less than 10 and tolerances greater than 0.10, indicating that our data do not suffer from multicollinearity issues.

5 Analysis and Results

Multi-level structural equation modeling (MSEM) with Mplus7.4 (v) [44] was employed to test proposed relationships. By following the guideline of Hofmann and Gavin [45], we group-mean centered all the individual level variable (i.e., rated by leader such as team creativity and employee creativity). We performed parameter based approach to estimate the confidence intervals (CIs) for the direct and indirect path [46].

5.1 Preliminary Analysis

Although our hypotheses represent individual levels within a group, we examined both within and between variances to ensure whether data is partitioned into within-group only or between groups as well. We did not find any variable falling under between levels and all between level variances among the construct were insignificant. It implies that our data is nested within group level which is also aligned with our proposed hypotheses. Next, Since SMKDG, knowledge utilization, knowledge elaboration and knowledge leadership captured at team-level, therefore individual-level data was aggregated to the team-level [47].

First, we calculated interrater-agreement (i.e., rwg-uniform and rg-slightly skewness), interclass correlation (i.e., ICC1), reliability of group mean (i.e., ICC2), and between the group variance (i.e., F statistic). Analysis showed that Cronbach's alpha, interrater-agreement are pretty good and in line with the cut-off values (>.70) [48]. Table. 1 present the within-group agreements and reliabilities.

Table 1. Within group agreement and reliabilities

Construct	Cronbach's alpha	$r_{wg-uniform}$	$r_{wg-slightly\ skewness}$	ICC_1	ICC_2	F	p. value
SMKDG	.83	.85	.81	.34	.68	3.45	.01
Knowledge utilization	.90	.87	.87	.25	.70	2.74	.01
Information elaboration	.87	.95	.94	.32	.79	1.64	.01
Employee creativity	.91	NA	NA	NA	NA	NA	NA
Team creativity	.94	NA	NA	NA	NA	NA	NA
KOL	.89	.88	.83	.41	.72	2.55	.01

Note: SMKDG = social media-based discussion groups, KOL = knowledge-oriented leadership.
Note: NA = not applicable (because of Individual level construct).

5.2 Confirmatory Factor Analysis (CFA)

Confirmatory factor analysis (CFA) was performed using Mplus8.0 software to compare the fit indices of the baseline model (six-factor model) with those of the competing models (five-, four-, three-, two-, and single-factor models) (see Table 2 and 3). Table 2 shows the results of the CFA indicated that the fit indices of the baseline model were significantly better than those of the other competing models ($\chi^2/df = 1.35$, RMSEA = 0.05, CFI = 0.95, TLI = 0.94, SRMR = 0.04), indicating that the variables have good structural validity.

Table 2. Confirmatory factor analysis and reliabilities

Construct	CFA range	CR	AVE
SMKDG	.87–.75	.92	.67
Knowledge utilization	.90–.81	.87	.87
Information elaboration	.87–.77	.91	.84
Employee creativity	.91–.82	.91	.85
Team creativity	.94–.74	.92	.79
KOL	.89–.82	.88	.83

Note: SMKDG = social media-based discussion groups, KOL = knowledge-oriented leadership.
Note: NA = not applicable (because of Individual level construct).

Table 3. Series of factor analysis and model fit indices

Model	χ^2/df	RMSEA	IFI	CFI	TLI	SMR
Six-factor (SMKDG + KU + IE + EC + TC + KOL)	1.35	.05	.95	.99	.94	.04
Five-factor model (KU + IE + EC + TC + KOL)	1.98	.08	.89	.88	.84	.07
Four-factor model (KU + IE + TC + KOL)	3.95	.12	.74	.75	.72	.12
Three-factor model (SMKDG + KU + IE)	5.87	.25	.65	.66	.61	.24

Note1: SMKDG = social media-based discussion groups, KOL = knowledge-oriented leadership, KU = knowledge utilization, EL = information elaboration = employee creativity, TC = team creativity, Note2: NA = not applicable (because of Individual level construct), RMSEA = root mean squate approximation, IFI = incremental fit-index, comparative-fit index, Tucker-Lewis fit index, SMR = square multiple correlation

6 Hypotheses Testing

In line with the proposed hypotheses, the standardized results indicate the knowledge utilization is positively mediate the effect of SMKDG with on team creativity (*Indirect effect* = .05, 95% *CI* = [.01, .19]), and (*Indirect effect* = .09, 95% *CI* = [.03, .12]), and information elaboration on employee creativity (*Indirect effect* = .23, 95% CI = [.10, .35]) respectively, hence H1 and H2 are supported. Finally, we tested moderated-mediation effect of KOL and found that the mediating effect of at higher levels of knowledge leadership (β = 0.39, p < 0.001), 95% *CI* [0.29, 0.54]) was significantly higher than that of lower levels of KOL (β = 0.22, p < 0.001, 95% CI is [0.14, 0.38]. The results suggest that KOL positively moderate the mediating effect of knowledge utilization thereby supporting H3. Likewise, at higher levels of KOL (β = 0.23, p < 0.05), 95% CI [0.04, 0.26]) mediating effect was more significant compared with lower levels of KOL (β = 0.17, p < 0.01, 95%) CI was [0.09, 0.35]. The results suggest that KOL positively moderate the mediating effect of information elaboration thereby supporting H4.

7 Discussion

The current study provides an understanding of predictors of creativity that may provide companies with valuable insights on how to facilitate knowledge utilization and information elaboration by creating SMKDGs and promoting KOLs in organizations.

To address our first RQ1, we tested the mediating effects of knowledge utilization and knowledge elaboration between SMKDGs and creativity at the team and employee levels. As expected, the results support our proposed hypothesis (i.e., H1 and H2) and confirm the above-mentioned mediation mechanism. This means that SMKDGs have the potential to help members share knowledge and information, enabling them to access and utilize knowledge resources and information easily and effectively, which in turn reflects increased creativity across the organization. This is consistent with existing research in other contexts, confirming that knowledge utilization is aggregated into task-related knowledge bases in teams, and sufficient knowledge resources are reserved for team innovation, so that teams can achieve creative collision through the integration of rich knowledge elements [10, 49]. Likewise, at the individual level, employees can search for a large amount of information from colleagues through SMKDGs [7], and then integrate their own information with that of colleagues to process tasks. In the process of information elaboration, the creativity of employees will be enhanced.

To address the second RQ2, we examined the moderating mediation of KOLs in contextualizing the effects of SMKDGs on creativity-promoting. The results confirmed that KOL positively moderated the mediating effect of team knowledge utilization. Drawing on KBV, respect for knowledge and knowledge workers in the context of high knowledge leadership helps to shape a knowledge management system where everyone is willing to contribute knowledge. As a result, team members are more willing and able to collaborate, share and leverage knowledge through TSMU. This centripetal organizational climate is ultimately reflected in increased creativity. High-level KOLs can create a positive team learning atmosphere and promote active learning among employees. When teams collaborate with social media, employees are more proactive in exhibiting knowledge behaviors, which in turn reinforces the accumulation of tacit knowledge at the individual level, allowing them to process heterogeneous and diverse information more effectively, thereby completing the knowledge creation spiral.

8 Theoretical Implications

We summarized theoretical contribution as follows: First, despite the increased research interest in social media and creativity, minimal research has provided empirical evidence on the relationship between SMKDGs and different levels of creativity. Extant literature provides tranches of social media-based KM systems, leadership, and creativity in isolation. Our study links theoretical and validate empirically KM and leadership literature, and pave the way for knowledge-intensive industries to foster creativity through SMKDGs and KOLs.

Second, prior research on the mechanism of TSMU on creativity mostly focused on the knowledge sharing perspective [7]. However, KM not only including knowledge sharing, but also containing knowledge application [39, 49]. Our study explores the ways such as knowledge utilization and knowledge elaboration from a multi-level perspective to promote creativity. These mechanism helps provided thoughtful insight on how SMKDG are positively influence the creativity of teams and employees.

Third, most previous studies have used knowledge leadership as a predictor variable of creativity to explore the mechanism of knowledge leadership on creativity [50]. However, given the nature of the research context, the current study discusses a unique type of leadership behavior within the KM initiative, known as KOL [51]Our study thus contributes to the existing leadership literature supporting creativity and knowledge transfer among employees embedded in social networks. Knowledge leadership is an essential contextual variable that can significantly accelerate creativity enhancement by clarifying organizational development with employees, creating a learning atmosphere, and acting as a role model.

9 Practical Implications

For managers, our research has important implications. First, organizations should take a holistic view of SMKMDGs and acknowledge their values. Managers should focus not only on the role of SMKMDGs in accomplishing tasks, but also on the emotional value that SMKMDGs bring within the organization. In other words, leaders need to help team members achieve dynamic, cross-regional communication and collaboration by building internal social networks to promote the effective role of SMKMDGs as channels for rapid communication, efficient collaboration, and resource sharing., our research may provide a mechanism to create a platform that facilitates real-time communication, creating an atmosphere of mutual trust in organizations [27].

Second, managers should use SMKMDGs as a team, focusing on cultivating the ability to generate new knowledge and realize new services using massive, scattered and fragmented information. By doing so, employees will be able to generate creativity in work-related tasks. This practice helps to create a culture that is conducive to the use of team knowledge and the interpretation of employee information, and enhances the creativity of the team and employees by integrating the maximum utility of knowledge [5].

Third, organizations should select KOLs through multiple channels, focus on recruiting talents, and change the way of matching a organization. Organization should focus on cultivating KOL's professional skills and cultural knowledge, improve their quality [8, 52], and provide enterprises with strong human resource management (HRM) support while KM.

10 Limitation and Future Research Directions

First, this study was conducted in China, ignoring differences in the impact of SMK-MDGs on creativity in other cultures. Future studies can test our research model in different geographic regions and introduce a diverse sample to enhance the robustness and external validity of the findings. Second, given the different nature of social media in China, future research should examine SMKMDGs in real work scenarios using other social media platforms, such as other private social media. Third, as we adopt the positive aspects of using social media for creativity, however, previous research has highlighted that social media-based discussions can lead to psychological shifts and interruption overload [7]. Future research should also consider other mediating mechanisms, such as overload perspective and how it affects creativity levels.

11 Conclusion

Although technological advancements in social media greatly support creativity and enable people to innovate with different thought processes. However, no studies have examined the impact of social media knowledge discussion groups (SMKDGs) on creativity through knowledge utilization, and the impact of SMKDGs on employee creativity through information elaboration in a multi-level model. In doing so, we respond to a recent call for research and a lively debate that emphasizes studying the impact of social interaction on creativity [53]. By adopting a KM perspective, the current study highlights a practical examination of how SMKDGs can play a role in promoting team and employee level creativity. The current study provides insights into the underlying mechanisms of how SMKDGs enrich employee cognitive processes to support information utilization and information elaboration. We find that statistics support the mediating role of knowledge utilization and information elaboration at two levels. We reassess the role of social media in building systems of social interaction and pave the way for organizational think tanks to use social media as a mechanism for creativity and sustainable human capital.

Furthermore, this study sheds light on the role of leadership, especially in knowledge-intensive industries, to foster creativity through SMKDGs. Unlike previous research focusing on traditional roles of leaders (e.g., exploitative leadership), we provide insights into leadership behavior in the context of knowledge management and can foster creativity in employees and teams. Finally, we examine the role of SMKDGs from a multi-level perspective and argue that SMKDGs have a positive impact on team and employee creativity when KOL interactions are high. We found that KOL support positively moderated the mediating effects of knowledge utilization and information elaboration at both levels. Our findings complement research on SMKDGs and creativity and have instructive value for teams to effectively use social.

Acknowledgments. The work is supported by the National Natural Science Foundation of China (Nos. 72171047 and 71771053), the Natural Science Foundation of Jiangsu Province (No. BK20201144), and the Key Project of Social Science Foundation of Jiangsu Province(21GLA002).

References

1. Luqman, A., et al.: Empirical investigation of Facebook discontinues usage intentions based on SOR paradigm. Comput. Hum. Behav. **70**, 544–555 (2017)

2. Teece, D.J.: Explicating dynamic capabilities: the nature and microfoundations of (sustainable) enterprise performance. Strateg. Manag. J. **28**(13), 1319–1350 (2007)
3. Leal-Rodríguez, A., et al.: Knowledge management and the effectiveness of innovation outcomes: the role of cultural barriers. Electron. J. Knowl. Manag. **11**(1), 62–71 (2013)
4. Leal-Rodríguez, A.L., et al.: Knowledge management, relational learning, and the effectiveness of innovation outcomes. Serv. Ind. J. **33**(13–14), 1294–1311 (2013)
5. Lindblom, J., Martins, J.T.: Knowledge transfer for R&D-sales cross-functional cooperation: unpacking the intersections between institutional expectations and human resource practices. Knowl. Process. Manag. **29**, 418–433 (2022)
6. Luqman, A., et al.: Untangling the role of power in knowledge sharing and job performance: the mediating role of discrete emotions. J. Knowl. Manag. (2022) (ahead-of-print)
7. Luqman, A., et al.: Does enterprise social media use promote employee creativity and well-being? J. Bus. Res. **131**, 40–54 (2021)
8. Luqman, A., et al.: Enterprise social media and cyber-slacking: an integrated perspective. Int. J. Hum.-Comput. Interact. **36**(15), 1426–1436 (2020)
9. Becker, D., Jüttner, K.: The impedance of fast charge transfer reactions on boron doped diamond electrodes. Electrochim. Acta **49**(1), 29–39 (2003)
10. Ali-Hassan, H., Nevo, D., Wade, M.: Linking dimensions of social media use to job performance: the role of social capital. J. Strateg. Inf. Syst. **24**(2), 65–89 (2015)
11. Nusrat, A., et al.: Enterprise social media and cyber-slacking: a Kahn's model perspective. Inf. Manag. **58**(1), 103405 (2021)
12. Saleem, S., Feng, Y., Luqman, A.: Excessive SNS use at work, technological conflicts and employee performance: a social-cognitive-behavioral perspective. Technol. Soc. **65**, 101584 (2021)
13. Richter, A.W., et al.: Creative self-efficacy and individual creativity in team contexts: cross-level interactions with team informational resources. J. Appl. Psychol. **97**(6), 1282 (2012)
14. Stahl, G.K., et al.: A look at the bright side of multicultural team diversity. Scand. J. Manag. **26**(4), 439–447 (2010)
15. Gong, Y., et al.: A multilevel model of team goal orientation, information exchange, and creativity. Acad. Manag. J. **56**(3), 827–851 (2013)
16. Tuan, L.T.: Organisational ambidexterity and supply chain agility: the mediating role of external knowledge sharing and moderating role of competitive intelligence. Int. J. Log. Res. Appl. **19**(6), 583–603 (2016)
17. Benitez, J., et al.: IT-enabled knowledge ambidexterity and innovation performance in small US firms: the moderator role of social media capability. Inf. Manag. **55**(1), 131–143 (2018)
18. Abubakar, A.M., et al.: Knowledge management, decision-making style and organizational performance. J. Innov. Knowl. **4**(2), 104–114 (2019)
19. Zulfiqar, S., Khan, Z., Huo, C.: Uncovering the effect of responsible leadership on employee creative behaviour: from the perspective of knowledge-based pathway. Kybernetes (2022). (ahead-of-print)
20. Nabi, M.N., Liu, Z., Hasan, N.: Investigating the effects of leaders' stewardship behavior on radical innovation: a mediating role of knowledge management dynamic capability and moderating role of environmental uncertainty. Manag. Res. Rev. (2022)
21. Shehzad, M.U., et al.: Knowledge management enablers and knowledge management processes: a direct and configurational approach to stimulate green innovation. Eur. J. Innov. Manag. (2022). (ahead-of-print)
22. Wang, Y., et al.: Study on the public psychological states and its related factors during the outbreak of coronavirus disease 2019 (COVID-19) in some regions of China. Psychol. Health Med. **26**(1), 13–22 (2021)
23. Amit, R., Schoemaker, P.J.: Strategic assets and organizational rent. Strateg. Manag. J. **14**(1), 33–46 (1993)

24. Kogut, B., Zander, U.: Knowledge of the firm, combinative capabilities, and the replication of technology. Organ. Sci. **3**(3), 383–397 (1992)
25. Zhou, G., Gul, R., Tufail, M.: Does servant leadership stimulate work engagement? The Moderating role of trust in the leader. Front. Psychol. **13**, 925732 (2022)
26. Al Khajeh, E.H.: Impact of leadership styles on organizational performance. J. Hum. Resour. Manag. Res. **2018**, 1–10 (2018)
27. Jackson, T., et al.: Managerial factors that influence the success of knowledge management systems: a systematic literature review. Knowl. Process. Manag. **27**(2), 77–92 (2020)
28. Nisar, T.M., Prabhakar, G., Strakova, L.: Social media information benefits, knowledge management and smart organizations. J. Bus. Res. **94**, 264–272 (2019)
29. Ram, J., Titarenko, R.: Using social media in project management: behavioral, cognitive, and environmental challenges. Proj. Manag. J. **53**(3), 236–256 (2022)
30. Alashwal, A.M., Abdul-Rahman, H., Radzi, J.: Knowledge utilization process in highway construction projects. J. Manag. Eng. **32**(4), 05016006 (2016)
31. Liu, G., Tsui, E., Kianto, A.: Revealing deeper relationships between knowledge management leadership and organisational performance: a meta-analytic study. Knowl. Manag. Res. Pract. **20**(2), 251–265 (2022)
32. Sigala, M., Chalkiti, K.: Knowledge management, social media and employee creativity. Int. J. Hosp. Manag. **45**, 44–58 (2015)
33. Lu, B., Yan, L., Chen, Z.: Perceived values, platform attachment and repurchase intention in on-demand service platforms: a cognition-affection-conation perspective. J. Retail. Consum. Serv. **67**, 103024 (2022)
34. Hoever, I.J., et al.: Fostering team creativity: perspective taking as key to unlocking diversity's potential. J. Appl. Psychol. **97**(5), 982 (2012)
35. Chen, M., et al.: Environmental cost control system of manufacturing enterprises using artificial intelligence based on value chain of circular economy. Enterp. Inf. Syst. **16**(8–9), 1856422 (2022)
36. Chaithanapat, P., et al.: Relationships among knowledge-oriented leadership, customer knowledge management, innovation quality and firm performance in SMEs. J. Innov. Knowl. **7**(1), 100162 (2022)
37. Masood, A., et al.: Linking enterprise social media use, trust and knowledge sharing: paradoxical roles of communication transparency and personal blogging. J. Knowl. Manag. (2022) (ahead-of-print)
38. Zhong, E., et al.: Comsoc: adaptive transfer of user behaviors over composite social network. In: Proceedings of the 18th ACM SIGKDD International Conference on Knowledge Discovery and Data Mining (2012)
39. Sung, S.Y., Choi, J.N.: Effects of team knowledge management on the creativity and financial performance of organizational teams. Organ. Behav. Hum. Decis. Process. **118**(1), 4–13 (2012)
40. Farmer, S.M., Tierney, P., Kung-McIntyre, K.: Employee creativity in Taiwan: an application of role identity theory. Acad. Manag. J. **46**(5), 618–630 (2003)
41. Shin, S.J., Zhou, J.: When is educational specialization heterogeneity related to creativity in research and development teams? Transformational leadership as a moderator. J. Appl. Psychol. **92**(6), 1709 (2007)
42. Podsakoff, P.M., et al.: Common method biases in behavioral research: a critical review of the literature and recommended remedies. J. Appl. Psychol. **88**(5), 879 (2003)
43. Talwar, M., et al.: Has financial attitude impacted the trading activity of retail investors during the COVID-19 pandemic? J. Retail. Consum. Serv. **58**, 102341 (2021)
44. Muthén, B., Muthén, L.: Mplus. Chapman and Hall/CRC, UK (2017)
45. Hofmann, D.A., Gavin, M.B.: Centering decisions in hierarchical linear models: implications for research in organizations. J. Manag. **24**(5), 623–641 (1998)

46. Bauer, D.J., Preacher, K.J., Gil, K.M.: Conceptualizing and testing random indirect effects and moderated mediation in multilevel models: new procedures and recommendations. Psychol. Methods **11**(2), 142 (2006)
47. James, L.R., Demaree, R.G., Wolf, G.: Estimating within-group interrater reliability with and without response bias. J. Appl. Psychol. **69**(1), 85 (1984)
48. Bliese, P.D.: Within-group agreement, non-independence, and reliability: implications for data aggregation and analysis (2000)
49. Lynch, J., West, D.C.: Agency creativity: teams and performance: a conceptual model links agency teams' knowledge utilization, agency creativity, and performance. J. Advert. Res. **57**(1), 67–81 (2017)
50. Donate, M.J., de Pablo, J.D.S.: The role of knowledge-oriented leadership in knowledge management practices and innovation. J. Bus. Res. **68**(2), 360–370 (2015)
51. Alavi, M., Leidner, D.E.: Knowledge management and knowledge management systems: conceptual foundations and research issues. MIS Q. **25**, 107–136 (2001)
52. Li, C.R., et al.: A multilevel model of team cultural diversity and creativity: the role of climate for inclusion. J. Creat. Behav. **51**(2), 163–179 (2017)
53. Khalid, J., et al.: After-hours work-related technology use and individuals' deviance: the role of other-initiated versus self-initiated interruptions. Inf. Technol. People (2021)

Online Training of Employees of the Enterprises in the Creative Economy: Benefits and Features

Nataliia Danylevych[1]([✉]) [iD], Svetlana Rudakova[1] [iD], Oksana Poplavska[1] [iD], Liudmyla Shchetinina[1] [iD], and Oleksandr Rudakov[2] [iD]

[1] Kyiv National Economic University named after Vadym Hetman, 54/1 Prospect Peremogy, Kyiv 03057, Ukraine
{danylevych.nataliia,oksana.poplavska}@kneu.edu.ua,
svetlana.rudakova.home@gmail.com, sludval@ukr.net
[2] "LINIA KINO" Ltd., Board Member of Ukrainian League of Industrialists and Entrepreneurs, 34B Predslavinska St., Kyiv 03150, Ukraine
orudakov@gmail.com

Abstract. The development of a new global ecosystem increases requirements for the continuous refreshing of knowledge and expanding the skills of all employees throughout their working lives. In the conditions of individualization and isolation of society, the expansion of the IT-technologies usage makes it possible to simplify the task of creating such a system and make the learning process accessible to every employee. The purpose of the article is to analyze the effectiveness of online training for personnel of companies in the creative sector of the economy of Ukraine. To evaluate the effectiveness of online learning, a survey was carried out among companies that belong to the creative economy of Ukraine. The purpose of the survey was to determine the level of satisfaction of the management with the results of employee training and development management system in their companies. The survey was conducted in parallel, from December 2021 to January 2022, participants were the representatives of the top management of companies and heads of HR departments; the selection of survey participants has been carried out automatically and was randomized. A total of 53 Ukrainian companies took part in the study. The research proves that online learning is no less effective than full-time learning, however, its effectiveness largely depends on the correct selection of programs and methods of staff training. We recommend the extended use of gamification for the sphere of the creative economy. The current level of IT-technology development makes it possible to combine full-time, mixed, and e-learning, enhancing the synergistic effect.

Keywords: E-learning · Creative economy · Express survey · Gamification · IT technologies

1 Introduction

The growing entropy in the modern world adds more challenges for the societies and economies of many countries. At the same time, the understanding of what is happening

J. Maślankowski et al. (Eds.): PLAIS EuroSymposium 2022, LNBIP 465, pp. 65–84, 2022.
https://doi.org/10.1007/978-3-031-23012-7_5

lies beyond technological breakthroughs or unaccounted-for factors such as pandemics. What is important is the emergence of a new socio-economic system where digital technology is playing a leading role. The ongoing development of a new global ecosystem implies an ever-growing need to acquire a lot of employees with the relevant knowledge and skills and to train staff throughout their working lives. Therefore, adapting the model of company management to modern challenges requires special attention and resources allocated for staff training.

The development of IT technologies in the conditions of individualization, isolation of societies, and global division of labor allows for simplifying and making the learning process available to everyone. Given the consequences of COVID-19, companies have become more active in the online learning environment, as it meets the needs of employees in a safe working environment and reduces the risk of workforce loss for the organizations. In fact, for business organizations, the development of online learning has helped to minimize financial costs, increase the effectiveness of training, and increase loyalty, productivity, and motivation of staff through career management and talent development. At the same time, the degree of effectiveness of the implementation of online training projects largely depends on the correctness of the tasks, definition of goals, and selection of tools and methods of learning. Adapting learning tools to the strategic goals of the company and choosing the most effective model of staff training is an important task for the business.

This study aims at studying the experience of online training of the staff of companies in the creative economy of Ukraine to identify problems and find the most acceptable practices for online training of staff in this sector of the economy.

The hypothesis is to confirm a stable correlation between the improved performance of the personnel of companies in the creative sector of the economy and the increase of the social effect in personnel management when using online learning, as well as the varying degree of influence of the selected online learning tools on the chosen indicators during the pandemic.

2   Literature Review

In the 21st century, technology has become an integral part of our daily lives and an unconditional requirement for doing business. A significant transformation in the approaches to learning by different groups of the population requires a rethinking of conceptual and practical solutions that contribute to effective changes in education. With the introduction of modern educational models in companies, the process of implementation of learning ideas throughout the career is happening due to those innovations. All of this requires the selection of relevant measures, the creation of structures, and the revision of learning technologies [1]. The business faces the need to organize a new educational process, which includes the acquisition of deep knowledge and professional competencies [2].

The importance of technology in education today is extremely pronounced. To create modern educational programs, it is necessary to carry out various types of research as well as to offer fundamentally new methods of delivering new information [3]. The use of online learning platforms is a new way of organizing the educational process, offers an alternative, and complements traditional methods of its organization. It creates opportunities for personal learning, group teaching, and interactive classes [4]. Data, analytics, artificial intelligence, and machine learning can also significantly affect adaptability and the ability to personalize educational technology [5–7].

The use of information technologies and the Internet in the education and training of personnel is one of the most important indicators of society's shift into the digital space, it will contribute to greater efficiency and effectiveness of the economy as a whole. The concept of training is no longer a traditional concept limited to the organization of training courses, but it has become a strategic choice in the system of investment and human resource development [8].

Modern education differs significantly from traditional teaching methods. Among the main features of modern education, distance and online education have become the quintessence of the learning process in the modern era [9]. Distance learning implies that the teacher and a trainee stay in different places, and it requires communication through technology and special institutions [10–12]. Thus people of all ages can develop their skills and expand their experience [13].

The education system while using new technological tools and unlimited information resources must effectively incorporate them into the educational process. The practice of online courses and mixed learning creates a field of endless educational opportunities that focuses on the quality of education for everyone regardless of their place of residence and skills, but according to their interests and capabilities [14].

Studying the concept of "Online Learning", the authors often call it "e-learning" as it is implemented in the Internet environment [15]. It is also called network education because the transfer of skills and knowledge to a large number of listeners at different times goes through computer networks [16]. At the same time, there is a dispute about whether online learning is one of the types of "distance learning". Delen E., Liew J. (2016) define online education as follows: it is the use of the Internet to obtain educational materials, interact with the material, teachers, and other students, as well as to receive assistance in the learning process to gain knowledge, build their own opinions, and to increase the level of experience [17].

Thus, online learning is a collaboration of approaches and principles of distance and offline learning with the use of IT technologies. Therefore, the model of online staff training should include the following components: (Fig. 1).

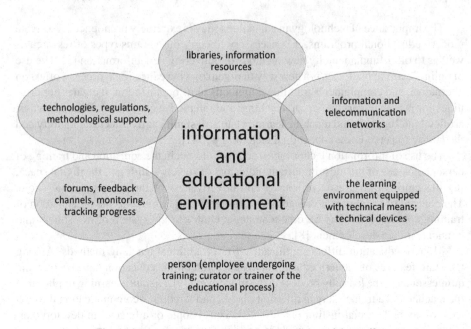

Fig. 1. The generalized model of online training of the staff.

Building an effective model of online learning in the company is impossible without understanding the nuances of adult learning. In principle, the adult learners tend to study better when education can be self-organized, the educational process is planned freely, classes include basic knowledge and practice, learning is active, solutions are relevant to current circumstances and can be applied to certain aspects of life [18]. Adult education has special functions, its development has content and formal specifics due to the characteristics of the people in need of training [19]. Psychological principles of adult learning are primarily related to the paradigm shift of modern education which focuses on development and self-development, creative initiative, independence, mobility, and competitiveness of future professionals. It is the psychological aspects of adult learning that allow us to build the most effective training models, and it is important for business companies which purchase training services or develop internal means for staff training to consider them [20].

Businesses need to accept that due to the development of information technology the idea of continuous education can be realized fully [21]. Participation in lifelong learning is crucial for navigating changing labor markets in an increasingly digital economy. This helps to maintain existing skills and acquire new ones. Even professionals with extensive experience need to improve their skills from time to time or acquire new skills that will help improve their position in the labor market. To achieve this various tools and types of online learning become useful. Online education is becoming the norm in our lives [22].

Given that the ongoing COVID-19 crisis has led to a significant increase in online adult learning, it is important to explore this area and identify the most acceptable learning models for the companies. However, the natural tendencies of many people to the reflection and procrastination should also be taken into account. The hardest thing to overcome the limitations of e-learning is resistance. E-learning will not be successful in a corporate culture that resists change. It is necessary to maintain the spirit of change while implementing online learning [23].

Employers should look for opportunities to organize training of their staff, and provide additional support to get the necessary training for employees going. Personal growth should be part of every job description [24]. Cooperation with private and state-owned enterprises plays an important role in maintaining digital transformation to address changes, adjustments, and updates to the educational programs and learning models [25].

The effectiveness of investment in the professional development of staff has long since been understood and is no longer doubted. But adult learning is always associated with change and transformation, so it can be effective only if it is provided with not only age and professional characteristics but also personal interests which are taken into account, and it needs to be based on partnership.

3 Methodological Approach

The effectiveness of e-learning was assessed in the companies of the creative economy of Ukraine which included a cinemas chain, some event management companies as well as retail and catering companies. Online training of company personnel took place on platforms - classroom.google.com, meet.google.com, zoom. The evaluation methods involved conducting a content analysis of regulatory documents followed by a SWOT analysis of the implementation of the online training of the staff, express survey of managers, and surveys of employees of companies that participated in the training. To do this, two forms were created and distributed online through corporate communication channels. A distinctive feature of the express survey was its purpose—to determine the satisfaction of the companies' management with the training management system in the company and the results of the employees' training. The survey and questionnaire were conducted in parallel regardless of the results obtained and took place from December 2021 to January 2022. The top management of companies and heads of HR departments took part in the express survey of executives; the selection of survey participants was carried out automatically and randomly. The structure of respondents and participants in the survey and questionnaire is given in Table 1. A total of 53 Ukrainian companies took part in the study.

Table 1. The profile of the respondents.

Profile of respondents	Total		Percentage	
	Express poll	Staff questionnaires	Express poll	Staff questionnaires
Gender				
Male	91	249	41.9	45.8
Female	126	295	58.1	54.2
Work in the company (by the number of employees)				
Large companies with more than 500 employees	62	183	28.6	33.6
Medium business (from 100 to 500 employees)	92	217	42.3	39.9
Small enterprises (up to 10 employees)	63	144	29.1	26.5
Online learning experience				
First-time learners	86	162	39.6	29.8
Studied before	131	382	60.4	70.2
Work in the company (by the types of businesses)				
Cinemas	29	82	13.4	15.1
Retail chains	67	198	30.9	36.4
Catering (bars, restaurants)	70	158	32.2	29.0
Event management companies	51	106	23.5	19.5

The study evaluated several indicators of the effectiveness of online training in companies: the share of employees trained in the company; increase in productivity of employees as compared to the period before training; satisfaction with the learning results by both managers and students, etc.

The indicator is evaluated by the value of the corresponding indicator. In this case, the indicator for the enterprise corresponds to the average value based on all the answers of the respondents of the studied enterprise.

Each indicator (I_i) was calculated by the formula:

$$I_i = \frac{[X_i - X_{max}]}{[X_{min} - X_{max}]},\qquad(1)$$

where X_i is the actual value of the estimated indicator for the company, which corresponds to a certain indicator;

X_{min}—the minimum achieved value of the indicator for all enterprises that are studied;

X_{max}—is the maximum achieved value of the indicator for all enterprises that are studied.

The integral efficiency factor *(FE)* of the online learning model of the company is calculated by the formula:

$$FE = \frac{\sum_{i=1}^{n} I_i}{n}, \tag{2}$$

where n is the number of indicators in the evaluation form.

Note that the model of online training of the company's personnel is a set of management decisions, organizational regulations, methodological approaches, technical means, and channels to achieve the goals. Therefore, understanding the effectiveness of the online training of the staff involves the degree of achievement of training objectives for the company. A scale of assessments with the appropriate factors is adopted:

Unsuitable level	Unsatisfactory level	Acceptable level	Sufficient level	Advanced level
0.00–0.20	0.21–0.40	0.41–0.60	0.61–0.80	0.81–1.00

Assessment of the effectiveness of the company's staff online training model and achieving the goals involve improving both economic and social performance.

4 Conducting the Research and the Results

The choice of indicators to assess the effectiveness of the companies is dictated by the main objectives which include economic benefits and an improvement of the employer's brand. Therefore, the sociological survey revealed the degree of achievement of social goals, while the survey of the internal documentation—economic indicators.

The results of the study showed a growing interest of companies in the training programs for their personnel. At the same time, the choice of the topics in training programs is not necessarily related to the direct responsibilities of the employees. According to the management of companies, the main goals of training programs are forming a cohesive team, improving communication skills, teaching the staff the basics of time management, and the ability to cope with stress. It is the need to increase the number of contacts between employees during isolation due to anti-pandemic restrictions which led to the growth of online training (Fig. 2). Among the leaders of the training programs was the psychological training, and its secondary task was to harmonize the psychological state of employees to avoid depression and motivational burnout.

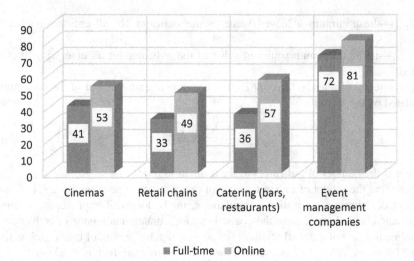

Fig. 2. Indicator 1—the share of staff who underwent training in the company during last year, %.

It should be noted that event management companies displayed the largest interest in various training programs (during the year 72% of employees took part in the full-time training and 81% in the online training). The least interest in training programs was shown by companies in the retail sector (only 33% of employees on average were full-time trained and 49% were trained online last year), which is due to the ability to master the necessary skills and company policies without extensive training.

It should be noted that online learning is cheaper than full-time one, but is still quite effective (Fig. 3). An economic evaluation of the benefits shows an increase in

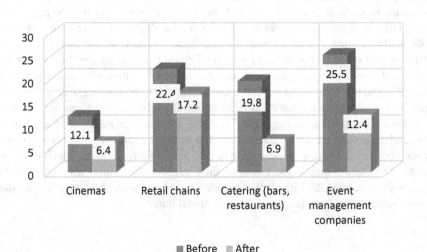

Fig. 3. Indicator 2—Return on Equity (ROE) (before and after the introduction of online training), %.

profitability for all groups of enterprises. At the same time, the greatest effect was noted among companies whose activities are related to catering (if for example for catering companies, the profitability of the full-time training project was at 6.9%, then online training Return on Equity (ROE) amounted to 19.8%, which is due to a significant reduction in travel expenses and other travel-related costs). In the context of increasing loss of profits due to the anti-epidemic constraints, online training has allowed reducing the cost of training while maintaining high efficiency.

There are also obvious benefits for other types of companies. The event management companies and cinemas by moving to online learning were able to double the profitability of programs by reducing the cost of training, while retail chains increased their profitability by 30%.

Such results were achieved by the growth of employees' productivity of work. At the same time, the comparison of the productivity of work depending on the form of education revealed no significant variations (Fig. 4).

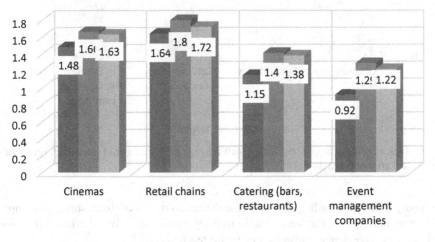

Fig. 4. Indicator 3—changes in productivity of the staff (before and after full-time and online training), US Dollars/Person.

As we can see from Fig. 4, regardless of the form of training, the productivity of workers is growing. For example, retailers report productivity growth of 10% for full-time training and 5% for online training. For the event management companies, productivity growth was 40% for the full-time training and 30% for online training. On average, all companies saw a 21% increase in productivity in full-time training and a 16% increase in online training. This difference is not significant but speaks of staff fatigue from online learning and the need for "live" communication.

No less important indicator of the effectiveness of training is the creativity level of the staff, which is defined as the changes in the number of innovations and proposals to improve processes in the company, the range, and quality of products and services. According to the study's results after undergoing online training, employees were more active in improving companies' work (Fig. 5).

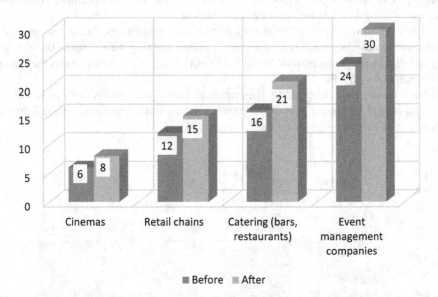

Fig. 5. Indicator 4—creativity level of the staff (change in the number of innovations/proposals (before and after online training).

Thus, online training helped to increase the creativity of staff in all surveyed companies, especially in cinemas—by 33%, in catering businesses—by 31%; in retail chains and event management companies, this figure reached—25%.

The introduction of the innovations allowed companies to improve their work and reduce the number of complaints and grievances from customers (Fig. 6). Thus, the number of complaints in cinemas has halved, and in event management companies it fell by 1.7 times, while in the retail and public catering businesses the fall was almost 50%.

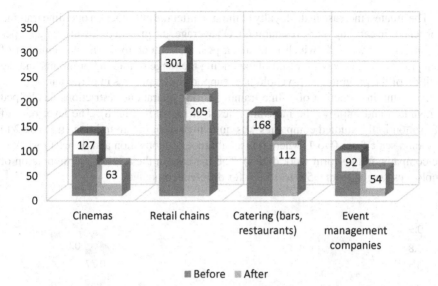

Fig. 6. Indicator 5—reduction in the number of claims/complaints/faults (before and after online training).

The social effect and result for companies during the implementation of online training can be regarded as the development of the companies' brand, and achieving high loyalty of the personnel is the defining goal. Using a rating scale from 0 to 1, where 0—no loyalty at all, and 1—maximum loyalty, the survey showed an increase in the loyalty of the staff after the introduction of online training (Fig. 7). At the same time, the maximum loyalty to the company was shown by cinemas chain employees (before training the loyalty rate was 0.62, and after training, it reached 0.78).

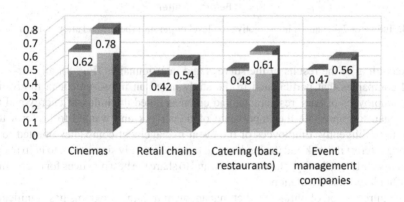

Fig. 7. Indicator 6—changes in the loyalty of the staff (before and after online training).

The greatest increase in the loyalty of the staff after the introduction of online training is observed in catering (bars, restaurants). On average, the growth of loyalty for all types of companies was 25%, which indicates a positive attitude to the implementation of such practices. This was the result of expanding the range of training programs and the addition of the programs of psychological support for employees in particular.

Also, the introduction of online training during quarantine restrictions has helped to maintain and improve the motivation level of the staff (Fig. 8). The level of staff motivation in the studied companies was quite high even before training (from 0.52 to 0.77 on a scale from 0 to 1, where 0 is an absence of motivation and desire to work in the company and perform work duties). The increase in the level of the motivation of employees ranged from 9.5% to 15.6% for different companies.

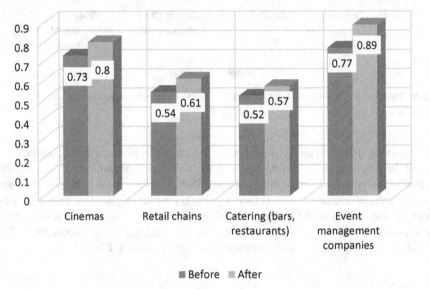

Fig. 8. Indicator 7—changes in the motivation level of the staff (before and after online training).

According to the results of the surveys and questionnaires (Fig. 8), all employees of all companies that participated in the study noted an improvement in motivation. Thus, employees of bars, restaurants, and cinemas noted the influx of energy and that they began to appreciate the importance of their work and wanted to perform their tasks more diligently. Employees of the event management companies pointed to the growing interest in work and the tendency to become socially significant, to help people, especially children, through their activities, and to share the best practices for overcoming the difficulties of self-isolation.

The introduction of online conflict management training programs has significantly increased productivity and the motivation of the staff. The results of the assessment of the level of conflict showed (according to the sociometry and the results of the survey) that after online training, employees became more confident and the number of conflicts decreased (Fig. 9).

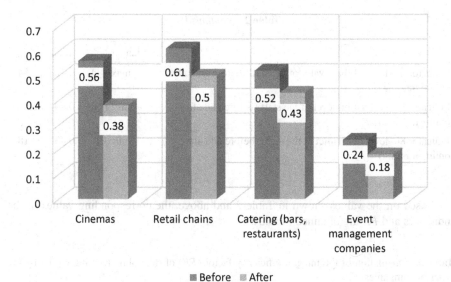

Fig. 9. Indicator 8—changes in the level of conflict in the team (before and after online training)

Figure 9 shows a significant reduction in the level of conflict between workers in the cinemas and event management companies (by 32.1% and 25% respectively). Smaller results are observed in the retail and catering companies (by 18% and 17.3%, respectively), which largely can be explained by the insufficient attention of the companies' management to this issue and the possibility of low effectiveness of the selected training courses to improve the social and psychological climate of the teams.

To calculate the integral efficiency factor (*FE*) of the online learning model, the following source data required to perform these calculations should be taken into account (Table 2).

Table 2. The source data for the integral efficiency of the online learning model calculation.

Indicator	Standard	X_{max}	X_{min}
Indicator 1—the share of staff who underwent training in the company during last year, %	→ max (100)	81	49
Indicator 2—ROE of training (before and after the implementation of online training)	→ max	25.5	12.1
Indicator 3—productivity of the staff (before and after training, including remotely)	→ max	1.72	1.22
Indicator 4—creativity level of the staff (changes in the number of innovations/proposals) (before and after online training)	→ max (100)	30	8
Indicator 5—number of claims/complaints/faults (before and after online training)	→ min (0)	205	54

(*continued*)

Table 2. (*continued*)

Indicator	Standard	X_{max}	X_{min}
Indicator 6—level of the loyalty of the staff (before and after online training)	→ max (≥1)	0.78	0.54
Indicator 7—motivation level of the staff (before and after online training)	→ max (≥1)	0.89	0.57
Indicator 8—level of conflict in the team (before and after online training)	→ min (0)	0.43	0.18

Based on the values shown in Table 2 and above, the corresponding values of the indicators and *FE* are obtained (Table 3).

Table 3. Calculation of the integral efficiency factor (*FE*) of the online learning model by the types of companies.

Indicator	Groups of companies			
	Cinemas	Retail chains	Catering (bars, restaurants)	Event management companies
Indicator 1—the share of staff who underwent training in the company during the year, %	0.875	1	0.75	0
Indicator 2—ROE of training (before and after the introduction of online training)	1	0.2313	0.4254	0
Indicator 3—productivity of the staff (before and after full-time and online training)	0.18	0	0.68	1
Indicator 4—creativity level of the staff (change in the number of innovations/proposals) (before and after online training)	1	0.6818	0.4091	0

(*continued*)

Table 3. (*continued*)

Indicator	Groups of companies			
	Cinemas	Retail chains	Catering (bars, restaurants)	Event management companies
Indicator 5—number of claims/complaints/faults (before and after online training)	0.9404	0	0.6159	1
Indicator 6—level of the loyalty of the staff (before and after online training)	0	1	0.7083	0.9167
Indicator 7—motivation level of the staff (before and after online training)	0.2813	0.875	1	0
Indicator 8—level of conflict in the team (before and after online training)	0.375	0	0.2188	1
Integral efficiency factor (FE) of the online learning model	*0.5815*	*0.4735*	*0.6009*	*0.4896*

According to the results of the calculations for all groups of companies, the introduced models of online training are[1] acceptable, which indicates the difficulties of project implementation, possible errors in the selection of the training methods, the unpreparedness of the staff for intensive and continuous training. Note that the effectiveness of the online learning programs for all companies will depend entirely on the internal policy and is developed according to the learning model, which, as the data in Table 3 shows, has a significant potential for improvement. In fact, most companies were mostly, up to 70%, focused on the online training, as per internal training records (training and retraining units), which likely narrowed the opportunities to acquire the necessary competencies, and limited the inclusion of relevant proposals; incorrect method of a training service provider selection is also probable (Table 4).

After reviewing the online training practices of all companies, the "gray areas" of the online staff training models were identified and summarized: in part, company management failed to restructure their staff training policies and used full-time learning programs; little attention was paid to the needs of the staff in social contacts and psychological support during online training, there were failures in communication between

---

[1] The online learning models are divided according to the type of educational service provider, i.e. the following models can be defined: all learning requirements are covered by the company's internal reserves; model that involves educational service providers only from the external educational market; and a combined model, which allows involving external resources while still using local resources and internal reserves.

Table 4. SWOT-analysis (generalized) of the effectiveness of the implemented practices of online training of the staff.

Strengths	Weaknesses
Qualified staff capable of conducting professional development training; Good equipment in the workplace which allows for implementing online training;	Lack of competent professional psychologists in companies; Lack of agreements on psychological support with third-party organizations; Lack of a clear educational strategy of the company; Incorrect criteria selected for the effectiveness of online learning
Opportunities	**Threats**
The motivation of staff to learn; Revision of work schedules and introduction of mixed work regulations (allowing the combination of work and study); Revision of training programs Revision of methods for selecting training service providers	Lack of necessary training programs on the market; The high cost of some relevant training programs; Inconsistency of teaching methods with the stated goals

the staff and management during training, the issues of the regulation of duties and work hours schedules were not resolved. Respondents also noted that some programs were boring and delivered poorly. All these can be solved by changing the approaches to the organization of labor processes and the proper selection of training methods.

5 Conclusion and Discussion

Research shows that online learning is no less effective than the full-time one, but its effectiveness largely depends on the correct selection of training programs and methods of the training of staff according to the specifics of work.

Given the dynamic development of educational services, it seems most acceptable to introduce gamification into the creative economy more widely. It should be noted that the gamification format differs from others games in that the participants are striving to achieve the goal of their real work. Game elements are integrated into real situations to motivate staff in the given conditions. The use of gamification is a convenient method of increasing the efficiency of the work because it destroys the boring routine. Let's look through the stages of implementation of gamification (Table 5).

The gamification makes it possible to quickly integrate employees into the work process, facilitate their involvement with the team and teach them the values and policies of the company. The gamification provides many opportunities for the training of the staff. Starting with the situational exercises in the form of games, which allow you to gain experience by simulating real situations, and ending with the learning in virtual reality. In addition, gamified training courses accelerate the process of a novice turning into an experienced worker.

Table 5. Stages of implementation of gamification for the staff training in the creative economy.

The name of the stage	Characteristics of the component
Defining goals	This means implementation of the special purposes gamified systems, for example, to increase indicator of the retention of customers and formation of commitment to the brand, or to increase the productivity of the work of employees. It is necessary to define them as accurately and concretely as possible, but leave the possibility to expand the main list, as it will need to be adjusted over time
Description of the desired behavior of the staff	The conduct and quantitative indicators of the best companies are collected together. After determining all types of the desired conduct, it is necessary to determine how to evaluate its progress. The whole gamification process is based on software-implemented algorithms. They invisibly convert all actions into figures, and then use these digitized data to create feedback. The figures may or may not be obvious to the players
Description of users and participants	It is necessary to determine which tools will be more appropriate for the goals and functions of the company. Segmentation is a standard practice in personnel management. Because games and gamified systems typically offer players a choice, they don't need to be limited to one target segment. Gamified systems can simultaneously include different actions for different groups of employees
Development of activity cycles	The game always has a beginning and an end, but during this time the game involves many cycles and branches to other actions. It is best to simulate the action in any gamified system with the help of activity cycles—a concept that is becoming popular in the description of social networks and social online activities. Solving the problem provokes some other actions, which in turn leads to the action of another player and so on. There are two types of cycles in the systems: cycles of attraction and promotion

(continued)

Table 5. (*continued*)

The name of the stage	Characteristics of the component
Introduction of game techniques	Combining all of the game elements and analyzing the players, goals, and motivation, it is easy to forget about the entertainment element of the system. When employees view the gamified system as a game, they are more likely to attach stronger to the work and carry out tasks more efficiently. It is necessary to regularly analyze the players' satisfaction with the system
Using appropriate tools	To properly implement gamification, you need a team of professionals with various skills. Necessary "tools": specialists who understand the business goals of the project; Game developers; Analysts who can understand the data generated by gamified systems; Technicians who know how to implement the vision of management

Developed by authors based on [26].

In conclusion, online learning technologies, and gamification, in particular, can be considered as a synergistic technology that allows using of the benefits of full-time, blended, and e-learning more efficiently, and can compensate and make negligible all of their shortcomings.

References

1. Biolcheva, P.: Trends in modern education. In: Conference: 2nd International Scientific Conference: Economics and Management (2018). https://doi.org/10.31410/EMAN.2018.838
2. Winter, E., Costello, A., O'Brien, M., Hickey, G.: Teachers' use of technology and the impact of Covid-19. Irish Educ. Stud. **40**(2), 235–246 (2021). https://doi.org/10.1080/03323315.2021.1916559
3. Paudel, P.: Online education: benefits, challenges and strategies during and afterCOVID-19 in higher education. Int. J. Stud. Educ. (IJonSE) **3**(2), 70–85 (2021). https://doi.org/10.46328/ijons
4. Poplavska, O., Danylevych, N., Rudakova, S., Shchetinina, L.: Distance technologies in sustainable education: the case of Ukraine during the coronavirus pandemic. In: The E3S Web of Conferences, p. 255 (2021). http://dx.doi.org/10.1088/1742-6596/1840/1/012050
5. Lodge, J.M., Kennedy, G., Lockyer, L.: Digital learning environments, the science of learning and therelationship between the teacher and the learner. In: Carroll, A., Cunnington, R., Nugent, A. (eds.) Learning under the Lens: Applying Findings from the Science of Learning to the Classroom. CRC Press, Abingdon (2020)
6. Tatomyr, I., Kvasnii, Z.: Artificial intelligence as a basis for the development of the digital economy: textbook. OKTAN PRINT, Praha, 376 p. (2021). https://doi.org/10.46489/aiabftd-07

7. Danylevych, N., Rudakova, S., Shchetinina, L., Poplavska, O.: Digitalization of personnel management processes: reserves for using chatbots. In: II International Scientific Symposium «Intelligent Solutions» IntSol-2021, September 28–30, 2021, Kyiv-Uzhhorod, Ukraine, pp. 166–176 (2021). http://ceur-ws.org/Vol-3106/Paper_15.pdf. Accessed 21 June 2022
8. Ben Amara, N., Atia, L.: E-training and its role in human resources development. Glob. J. Hum. Resour. Manag. **4**(1), 1–12 (2016)
9. Sun, A., Chen, X.: Online education and its effective practice: a research review. J. Inf. Technol. Educ.: Res. **15**, 157–190 (2016). https://doi.org/10.28945/3502
10. Dabbagh, N.: The online learner: characteristics and pedagogical implications. Contemp. Issues Technol. Teach. Educ. **7**(3), 217–226 (2007)
11. Kennepohl, D.: Accessible Elements: Teaching Science Online and at a Distance. Athabasca University Press, Athabasca (2010)
12. Holmberg, B.: The evolution, principles and practices of distance education. Bibliotheks und informationssystem der Univ, Oldenburg (2005)
13. Modern Education: A Significant Leap Forward. https://leverageedu.com/blog/modern-education/. Accessed 21 June 2022
14. Radziievska, O.H.: Information literacy and digital inequality: securing the child in the modern information space. Informatsiia i pravo (2017). http://ippi.org.ua/sites/default/files/12_3.pdf7. Accessed 21 June 2022
15. Stern, J.: Introduction to Online Teaching and Learning (2005). https://www.wlac.edu/online/documents/otl.pdf. Accessed 21 June 2022
16. What is 'E-learning'. https://economictimes.indiatimes.com/definition/e-learning. Accessed 21 June 2022
17. Delen, E., Liew, J.: The use of interactive environments to promote self-regulation in online learning: a literature review. Eur. J. Contemp. Educ. **15**(1), 24–33 (2016)
18. Principles of adult learning. By Stephen Lieb Senior Technical Writer and Planner, Arizona Department of Health Services and part-time Instructor, South Mountain Community College from VISION, Fall (1991). http://honolulu.hawaii.edu/intranet/committees/FacDevCom/guidebk/teachtip/adults-2.htm. Accessed 21 June 2022
19. Lukyanova, L.B.: The concept of adult education in Ukraine. Nizhyn, 24 p. (2011)
20. Smulson, M.L.: Student activity in virtual educational space. Virtual educational space: psychological problems (Psychology of the new millennium): international. Science-practice. Internet conference (28 May 2012). http://www.psy-science.com.ua/Konferenciya_2012_05_28/Smulson_Maryna_2012.doc. Accessed 21 June 2022
21. Pidhorodetska, I., Zozuliak-Sluchyk, R., Averina, K., Tykhonenko, O., Luchkevych, V., Karikov, S.: Informatização da educação como tendência da atividade educacional moderna. Laplage Em Revista **7**(3C), 494–499 (2021). https://doi.org/10.24115/S2446-6220202173C1650p.494-499
22. Semenikhina, O.V., Drushlyak, M.G., Bondarenko, Y.A., Kondratiuk, S.M., Ionova, I.M.: Open educational resources as a trend of modern education. In: 2019 42nd International Convention on Information and Communication Technology, Electronics and Microelectronics, MIPRO 2019 – Proceedings, pp. 779–782 (2019). https://doi.org/10.23919/MIPRO.2019.8756837
23. Ellis, P.F., Kuznia, K.D.: Corporate eLearning impact on employees. Glob. J. Bus. Res. **8**(4), 1–16 (2014). https://www.theibfr.com/download/gjbr/2014-gjbr/gjbr-v8n4-2014/GJBR-V8N4-2014-1.pdf. Accessed 21 June 2022
24. Czarnecka, A., Daróczi, M.: E-learning as a method of employees' development and training. In: Management, Organizations and Society. Agroinform, Budapest, pp. 95–104 (2017)

25. Morze, N.V., Kucherovska, V.O.: Ways to design a digital educational environment for K-12 education. In: CTE 2020: 8th Workshop on Cloud Technologies in Education, December 18, 2020, Kryvyi Rih, Ukraine (2020). http://ceur-ws.org/Vol-2879/paper08.pdf. Accessed 21 June 2022
26. Werbach, K., Hunter, D.: Gamification Toolkit: Dynamics, Mechanics, and Components for the Win. Wharton Digital Press, Boston (2015)

Big Data, Internet of Things and Blockchain Technologies

Barriers to Implementing Big Data Analytics in Auditing

Vitória Viana da Silva and Selma Oliveira[✉]

Fluminense Federal University, Volta Redonda, Rio de Janeiro, Brazil
{victoriavs,selmaregina}@id.uff.br

Abstract. Big data analytics (BDA) is increasingly becoming a critical component of the business decision-making process. The purpose of this study is to highlight the barriers/inhibitors in the implementation of an BDA. The interest of researchers and practitioners in the subject is growing. However, the body of empirical research on the main barriers in the implementation of a BDA in the field of auditing is still an underexplored topic. The research in question has as main objective to address the main inhibitors (barriers) for the implementation of an analytical big data in the audit. Based on prestigious literature; the main points of difficulties in the implementation of a BDA system were raised. Using a Likert-type scalar matrix, data from Audit professionals in Brazil were collected. The results indicated that the main barriers in the implementation of the BDA concern the dynamic and digital capabilities of organizations. On the other hand, big data analytics capabilities are not significant in most of its dimensions. The main contributions of this study are: (a) it identifies the main barriers/inhibitors in the implementation of a BDA in the field of auditing in Brazil; (b) serves as a guide for managers in their decision making in choosing resources to implement BDA; (c) this research is original and fills a gap in the literature on big data management. Therefore, it advances the body of knowledge on BDA implementation inhibitors.

Keywords: Auditing · Analytical Big Data · Barriers/Inhibitors

1 Introduction

Auditing within large organizations is undoubtedly one of the most crucial points in a management process. Both internal and external audits have a very well defined role within a company, as they are key points to aid in decision-making and identification of possible risks and bottlenecks within a process chain. The step is only effective from the point that more and more information is collected and gathered to form a basis reliable enough to serve as an aid for decision making and mainly for identifying risks and bottlenecks within a company's processes. This task is difficult to the point that the information, without the correct technology and trained professionals, can be inaccurate and too variable since they are linked to internal and external processes to the company in constant mutation. Relying on audit methodologies without the use of variability and volume of data is to remain in an environment of shallow and subjective conclusions [1].

© The Author(s), under exclusive license to Springer Nature Switzerland AG 2022
J. Maślankowski et al. (Eds.): PLAIS EuroSymposium 2022, LNBIP 465, pp. 87–108, 2022.
https://doi.org/10.1007/978-3-031-23012-7_6

[2] argue that due to major changes in organizational dynamics, currently direc-
tors are increasingly using information generated by databases instead of trusting their
instincts and opinions, since they are based on human behavior, without information. In
addition, there is a tendency for failures in the management of a company to occur. [3]
highlights that the occurrence of fraud in organizations is directly related to the behavior
and beliefs of individuals, such as the auditor's opinion based only on isolated cases.
In this scenario, the use of a network of information rich in variability and veracity to
assist in administrative analyses, brought the need to create technological tools, such as
Big Data. But what is Big Data?

Based on a study carried out by the Business Software Alliance (BSA), a company
specialized in software, it is estimated that approximately 2.5 quintillion bytes of content currently exist in the world, in text or multimedia format spread over hundreds of
platforms, splitting into thousands of categories of information. In this way, the need
arises to implement Big Data to facilitate the handling of this vast variety of data. [4]
highlighted the tool in the ranking of the best strategic trends in technology in 2013 and
as one of the most important technologies to be used in the coming years. Marked as an
artifact of the modern individual, Big Data is a technological environment where everything can be measured, captured and digitally documented to be ultimately transformed
into data, a process that today is also known as datafication.

In times of technological advances where the vast majority of documents are digitized
aiming at a sustainable and safer performance, auditors have a greater difficulty when
needing to analyze the immeasurable amount of data that an organization produces
financially and not financially speaking. It is on this deficiency that Big Data is supported
to justify its need within this pillar of business management, using the tool, auditors can
combine access to a broad database and efficiency in the separation and organization of
information. According to [5], in their article "Financial fraud and big data analytics –
implications detection on auditors' use of fraud brainstorming session", with a broader
data network it is possible to compare information over time in order to identify anomalies
in a more simplified way and improving forecast models that aim to point out possible
fraud risks.

Taking these points into consideration, it is worth noting that using Big Data in
au-diting processes is still an issue under constant discussion, as it is a relatively new
sub-ject both in practice and in theory. Be that as it may, audit firms are declaring that
Big Data is an increasingly essential part of their assurance practice. Data analysis, new
technologies, and access to detailed industry information will combine to help auditors
better understand the business, identify risks and issues, and provide additional insights
[6].

According to [7], in the corporate environment there are some barriers to the implementation of Big Data, such as: the high value involved in the implementation, since the
treatment of vast data generates a cost both for the training of employee and for the use
of software and databases that support the demand for information; the difficulty of handling such complex data and the need for a new range of specific knowledge in this new
technology, given that there are few studies regarding the implementation of Big Data in
audit routines. There is also a lack of trained professionals and structured frameworks for
a process of this size; legal and ethical issues that continue without a structure to serve

as a basis for the use of the mechanism, starting from the point that data manipulation is a delicate point due to dangers such as data leakage, it is necessary to implement regulations so that the process is not used in a way that will be legally harmful to the entity, even aiming at its own security in relation to the competitive market, among many others that will be addressed in this present work. It is valid, then, to bring exactly these barriers as a research justification for this article, seeking to show which are the inhibitors of the implementation of a Big Data for auditing. The research aims to address the deficit in the number of studies on the subject with regard to identifying these inhibitors, to a greater or lesser degree, in the context of independent auditing. This study presents the following contributions: (i) it highlights the main barriers in the implementation of an Analytical Big Data; (ii) It presents the analytical capabilities of Big Data; (iii) presents the dynamic and digital capabilities of Big Data; (iv) advances the body of knowledge in implementing a BDA. In short, big data analysis with its efficient analytical techniques can contribute to the audit to discover hidden patterns, correlations, and other insights from big data. It brings significant cost advantages, enhances the performance of decision making, and creates new services to meet customers' needs [4]. This article is structured according to the following sections: Literature review, methodology, results and analyses, and finally conclusions.

2 Literature Review

2.1 Auditing and Its Role in Accounting

There are several cases of scandals involving large companies that have taken place in recent decades, examples such as Enron and Toshiba have marked the accounting world and redirected the focus to major issues such as veracity and reliability of the information that is collected within an organization. Therefore, the audit, both internal and external, is placed as one of the biggest pillars within a corporate structure, both for internal control of everything that is processed and for shareholder decision making. Since it is such an important tool within a company, how can this process still be flawed to the point of causing situations in which the information collected is not enough to identify all the bottlenecks of a process [5]. For decades, the audit has been used by scholars as the result of collecting various data related to both the operating environment of a company and the corporate world as a whole. The professional in the area must be multidisciplinary, understand the various areas in which information can be supported and in what ways it will be useful within a decision making between organizational management. The information must be polished so that it is believable and reliable enough so that nothing goes unnoticed in the "eyes" of accounting [7].

In this way, there is an audit deficit in the light of contemporary technologies. The professional tends to have a limitation with regard to the three dimensions: speed, trust and impartiality. These dimensions, when facilitated through new technologies, facilitate the solution of limitations or inhibitors, refining the audit, allowing for better results and helping in the way in which shareholders see the organization [5]. [8] points out that many financial frauds are evidenced through intentional falsification of data collected from financial networks, inappropriate application of accounting standards, the so-called "creative accounting", modification of earnings management, among other practices that cause information asymmetry. The literature highlights that financial fraud has brought tremendous losses to large companies over the years, translated into damages penalized by the legal framework and the market. The second is significantly greater, since reputational damage is currently a greater detriment to the company than just lawsuits in its history. In a study carried out by the US government accounting office, [9] shows that of the 585 companies investigated, the economic loss generated by reputational damage is 7.5 times greater than the losses caused by legal proceedings. In addition, shareholders lose interest in keeping their capital within a company with a history of corporate crimes even though the fear of investment tends to persist. Considering that two of the major factors that cause the loss of audit efficiency are the uncertainty and variability of professional errors, the process needs greater capacity and, mainly, technology.

2.2 Big Data and Data Management

Big Data tools favored possibilities with regard to large-scale data processing and Auditing is a promising field to be explored in light of its processes. Thus, in this way, we emphasize that technological tools take their place in the process as a refinement, generating greater reliability of information and consequently further basing the decision-making made from them. But what is Big Data? Marked as one of the trends in data processing of the last decade, the tool is brought to us in different forms and definitions. However, what remains unchanged is its relationship to the manipulation of varied, structured and complex data from financial and non-financial sources. [10] brings the concept of the 4 v's of Big Data and its information: Volume, Velocity, Variability and Veracity. Volume is the most easily understood concept since it is the one most attributed to the use of the tool, which is the amount of data it is capable of managing. Variety is already supported on the condition that data are presented in different ways and come from different sources. Therefore, it is possible to gather media of all existing formats, be it photo, text, audio, among others. Another important element is the speed at which this information is accessed and processed, thus facilitating work that would previously be done at the pace of a human being. Finally, we emphasize that the veracity of information promotes trust, cohesion and data accuracy. Therefore, Big Data is essentially the analysis of large and varied sets of data, from different sources and formats, whether they are structured or not, with the main objective of generating information [11]. For auditors, Big Data becomes useful in helping to inspect, clean and transform this data into favorable patterns to indicate possible failures in a process and even to assist in decision making, bringing evidence to the stakeholders of a company from the current internal and external scenarios to the business environment.

2.3 Barriers to Implementation

As it is a process still in its initial stage, facing several studies and improvements, it is important to highlight the main barriers of implementing the tool within an audit process. These difficulties are often the same as those found for the use of any other technology within an organizational environment [7]. The first is the lack of stakeholder confidence in financing new and expensive technology. Several researchers suggest that companies are reluctant to invest in this type of knowledge because it is a long-term and intangible investment. Therefore, the gains are not directly related to the implementation and the payback period is high [12]. Following the same logic, there are financial barriers. The budget of an organization itself, already suffers several variations due to high commercial and operational expenses, causing a difficulty in managing the company's capital and keeping the result within the desired. [13] indicate that funds for training in the IT area are among those that receive the least funding, hindering the adoption of information technologies in general. Big Data brings with it the need for advanced and qualified skills to manipulate your data, consequently increasing the cost, both for staff training and for technological input.

The complexity of the data can also be highlighted, combined with the concern with the security of this information and the legal issues that involve them. The need to implement data governance, according to [14], is essential because it is about how this data will be stored, analyzed and accessed and mainly to determine its value and relevance, making it an extremely complex process. From this, concerns about the reliability of the data obtained begin to emerge, unauthorized access can cost the company's reputation [9]. Finally, together with the aforementioned, legal and ethical barriers are one of the most critical. Today in Brazil, we have the LGPD (General Data Protection Law), which sets out a series of conditions for how the data of individuals must be treated and all its implications, according to Law n° 13709/2018. The concern with the collection and misuse of personal data, used for example in a survey of consumer profiles, must be strongly taken into account, since in recent years several companies have had to pay high fines for lack of technique with the personal data of its users. In this context, both the legal and ethical aspects bring harm to the image of a company, if there is no strong and structured compliance to prevent the transparency of the use of information from being damaged. The studies that can be used as support for the discussion of the topic are the ones mentioned below (Table 1).

Table 1. Publications on big data in the audit context

Author (s)	Title	Date of publication	Publication in journal
[15]	Correlates of the internal audit function's use of data analytics in the big data era: Global evidence	2021	Journal of International Accounting, Auditing and Taxation
[6]	Incorporating big data in audits: Identifying inhibitors and research agenda to address those inhibitors	2016	International Journal of Accounting Information Systems
[5]	Financial fraud detection and big data analytics – implications on auditors' use of fraud brainstorming session	2018	Managerial Auditing Journal

(*continued*)

Table 1. (*continued*)

Author (s)	Title	Date of publication	Publication in journal
[16]	Data analytics in auditing: Opportunities and challenges	2015	Business Horizons
[17]	Big data techniques in auditing research and practice: Current trends and future opportunities	2018	Journal of Accounting Literature
[18]	Big data analytics in financial statement audits	2015	Accounting Horizons
[19]	Motivation to use big data and big data analytics in external auditing	2019	Managerial Auditing Journal

3 Methodology

This research was based on the literature, in which barriers and/or bottlenecks in the implementation of an BDA were identified, not exclusively in the field of Auditing, but in general in organizations. Various bases were used to extract the articles: Web of Science, Science Direct, Emerald, Willey, etc. The following barriers were identified: the Analytical Capabilities of Big Data; Dynamic Capabilities and Digital Capabilities. In a second moment, a field research was carried out, using a scalar-type questionnaire, prepared based on the Capabilities: Big Data Analytics, Dynamics and Digital. The judgment matrix was prepared based on the study by [20, 2], which was adapted to meet the needs of the present research, in which each of the 28 components of the groups of barriers raised were classified in degrees from 0 to 5 - not significant or not relevant and 5 - very significant or greater relevance. Pre-tests were applied before final application. The form was prepared and sent via LinkedIn to professionals in the auditing area, focusing on Controller/Auditor positions. In addition, they were sent only to Brazilian profiles. In this way, companies are diversified into national and multinational companies, but all located in Brazil. Thus, 200 profiles were sent and 23 responses were returned, within a period of 1 month (June/2022 to July/2022). Data were organized in Tables (Excel) for the application of descriptive statistics techniques.

4 Results and Analysis

In this section, the results of the research referring to the groups of barriers identified in the implementation of the "BDA" are presented and analyzed. The results reflect the opinion of the auditors and treated using descriptive statistical techniques.

4.1 Barriers: Big Data Analytical Capabilities, Dynamic Capabilities, and Digital Capabilities

4.1.1 Big Data Analytical Capabilities

Group: Technological and Investments/Financial

The results presented in Fig. 1 highlight the averages of the results of the "Technological" group. This group concerns "access, visualization and sharing data and tools". Specifically, this group reports whether the company has access to large volumes of data and whether it has the ability to integrate internal data sources.

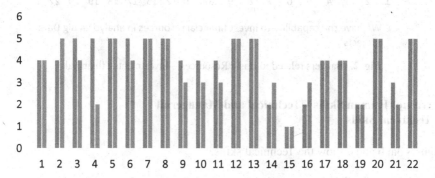

■ We have the capability to access large volumes of data

■ We have the capability to integrate multiple data sources for analysis

Fig. 1. Technological barriers - "Data"

On average, for most auditors there are substantial difficulties in accessing large volumes of data, since the responses are between grades 4 and 5, with 29.2% and 41.7% of the responses, respectively. On the other hand, on average, the ability to integrate multiple internal and external sources represents a moderately difficult barrier to implementing big data, with responses mostly concentrated between degrees 3 and 5. Both results converge to show how close the theory is to the theory. Practice. [20] suggest in their study that no matter how much a company is based on Information Technology, it is not enough to manage a BDA, since it is a more complex subject that requires more advanced technologies and skilled labor for its operation. Figure 2 shows the results of barriers related to the group "Basic Resources" – "Investments/Financial".

The results presented in Fig. 2 indicate (on average) that the capacity to invest in Big Data projects represents a substantial challenge for 38.09% of auditors. Making a parallel with the prestigious literature, the results obtained confirm the theories that companies, no matter how much they have capital for investment, are afraid to inject resources into BDA projects, thus creating a deficit in the implementation of the resource.

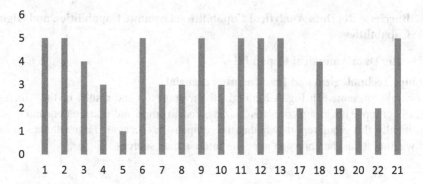

■ We have the capability to invest financial resources in analytical Big Data projects

Fig. 2. Barriers related to Basic Resources – Investments / financial.

- **Group: Human Skills – Technical and Managerial**
- **Technical Skills**

Figures 3a, b and c show the Technical skills.

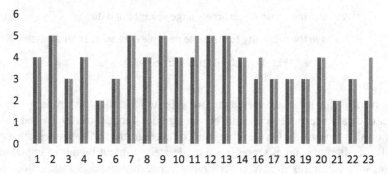

■ Our big data analytics team has the right skills to do their jobs successfully

■ Our big data analytics team is well trained to perform their duties

Fig. 3a. Barriers - Technical Skills"

By grouping together both answers to the questions: "Our BDA team has the right skills to do their jobs successfully" and "Our BDA team is well trained to perform their duties", it is possible to observe that the response pattern is, for the most part, the same. The professionals centered their answers between options 3 and 4, with a percentage of 29.2% and 33.3% respectively. These results indicate that there is a deficit both in training and in the skills needed to execute a BDA project.

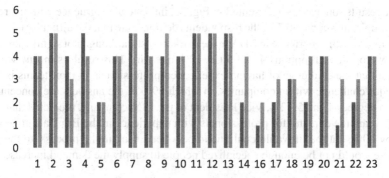

Fig. 3b. Barriers - "Technical" Skills

Training still represents a moderate challenge for companies (Fig. 3b). In the auditors' opinion, companies are still structuring themselves to provide training to professionals who will work on the front line with the big data analytical tool. On average, most responses (62.55%) are concentrated in degrees 3 and 4. However, it is important to note that 20.8% of responses are concentrated in degree 2. That is, for few auditors, there are companies that are already promoting training/qualification to their work teams. On the other hand, the answer to the question "Our BDA team is well trained", the results were concentrated in degrees 3, 4 and 5 (25%, 37.5% and 25% respectively), which indicates than moderate to strong results for the effectiveness of training. In summary, companies still have a long way to go in terms of training big data analytics teams.

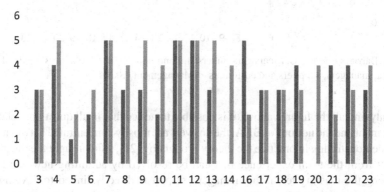

Fig. 3c. Barriers – "Technical" Skills

The results (on average) presented in Fig. 3c, indicate a congruence with the results of he group's questions, 30% of the respondents demonstrate that the difficulty of finding professionals with experience in BDA in the market exists, although not significant, while 43.4% were left with options 4 and 5 (21.18% in each). This result is in line with the question that the challenge of having experienced professionals in their teams is listed as a major challenge within companies, in which 69% of the answers are concentrated between degrees 4 and 5. [21], Vice President of the IBM company, predicted that in the coming years the demand for professionals with experience in the Big Data area would grow by 20%, but that it would continue to be a bottleneck in the process, to the point that there would not be enough specialized people to supply the demand increase.

- Management skills.

Figures 4a, b, c, e d highlight the results of managerial skills.

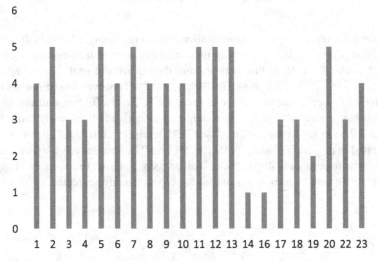

Fig. 4a. Barriers – Our BDA managers understand and appreciate the business needs of other functional managers, suppliers and customers - "Management" Skills

Analyzing all the figures above, it is possible to notice that in all questions related to management and the notions of BDA, the answers are mostly concentrated between 4 and 5, thus demonstrating in practice the theory brought by [20], who brought the theory that different from the team experience that can be resolved by admitting new employees or training the existing team, the management part is directly linked to development through the strong bond that must exist between company members, whether they from the same department or not. Thus, the deficit in this regard, according to the authors,

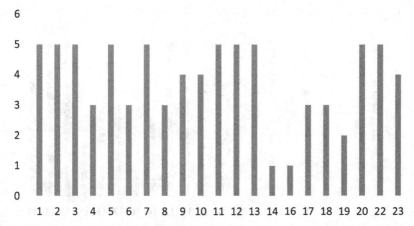

Fig. 4b. Barriers – Our BDA managers can work with functional managers, suppliers and customers to determine the opportunities big data can bring to our business - "Management" Skills

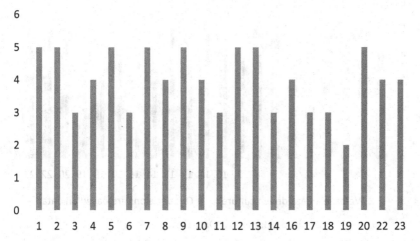

Fig. 4c. Barriers – Our BDA managers are able to anticipate the future suppliers of the future business needs of functional managers, and customers - "Management" Skills

means that the information extracted from Big Data can be done in an exemplary way and the use and application of this information will not be carried out in the most efficient way possible.

- Group: Data-Driven Culture (Intangible)

Figure 5 indicates the Data-Driven Culture group, considered a group of intangibles.

The data-driven culture concerns how companies use their data extraction and apply it in the company's day to day, such as in decision making. According to the chart above,

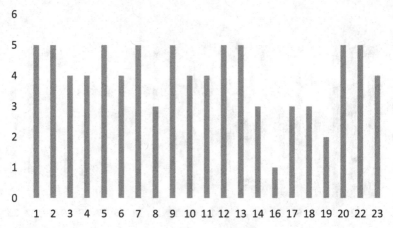

Fig. 4d. Barriers – Our BDA managers have a good sense of where to apply big data - "Management" Skills

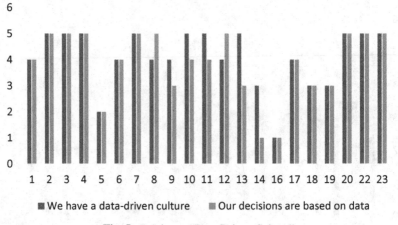

■ We have a data-driven culture   ■ Our decisions are based on data

Fig. 5. Barriers – "Data-Driven Culture"

78% of respondents indicate that having a data-driven culture and consequently having data-based decision-making is a recurring challenge within organizations. [22], through a survey carried out with 51 companies, found that although many companies invest and use BDA, few companies are able to extract value from the use of this tool, since they do not have a culture based on data.

• Organizational Learning Group (Intangible)

Figures 6a, b and c indicate the results of the group "Organizational Learning" (Intangible).

The questions addressed in the organizational learning group mostly had their answers within options 4 and 5, translating that it is still a constant and substantial

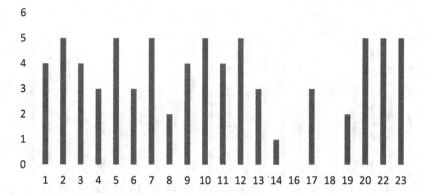

■ Do we share knowledge internally and externally with partners – stakeholders?

Fig. 6a. Barriers - "Organizational Learning"

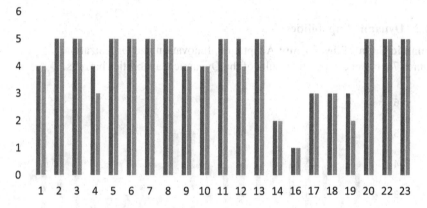

■ We have the capability to acquire new and relevant knowledge

■ We have the capability to explore new knowledge

Fig. 6b. Barriers - "Organizational Learning"

challenge for companies to be able to acquire and explore new and relevant knowledge. With 61% in the question "Do we share knowledge internally and externally with partners – stakeholders?", 64% in the questions in Fig. 2 and 73% in Fig. 3. Thus demonstrating that companies do not have a high capacity to use existing knowledge and of new ones so that the arrival of new technologies can be explored in a way that will add value in decision making in the most varied conditions that may arise [22].

■ We have the capability to absorb relevant knowledge

■ We have the capability to apply relevant knowledge

Fig. 6c. Barriers - "Organizational Learning"

4.1.2 Dynamic Capabilities

Dynamic Capabilities Group: Adaptation, Innovation and Organization.
Figures 7a, b and c show the results of the Dynamic Capabilities barriers.

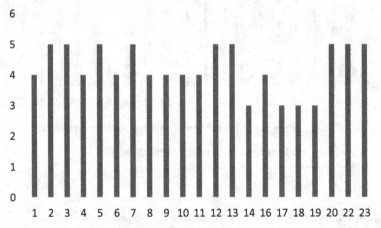

Fig. 7a. Barriers – We are capable of adapting/reconfiguring to changes - "Dynamic Capabilities – Adaptability"

Analyzing the group of Dynamic Capabilities (Figs. 7a, b and c), questions were created in order to understand the dynamism of companies in the face of changes and the main challenges for the organization's dynamics to contribute to the functioning of the BDA. From the results obtained in the three questions elaborated, we can conclude that, on average, for most respondents, companies have substantial difficulties to face the external environment in which they are inserted. Thus, the difficulties highlighted

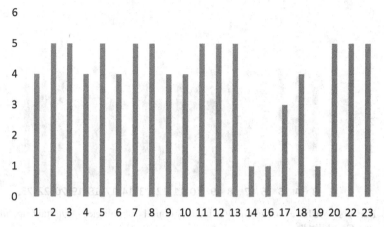

Fig. 7b. Barriers – We have the capacity for technological innovation - "Dynamic Capabilities – Capacity for Technological Innovation"

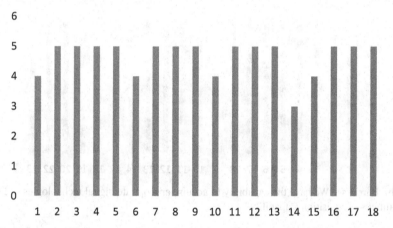

Fig. 7c. Barriers – We have organizational capacity - "Dynamic Capabilities – Organizational Capacity"

are: adaptation, innovation and organization. The responses of 72% of the auditors were mostly concentrated between grades 4 and 5, representing a significant challenge to be overcome by the companies under investigation.

4.1.3 Digital Capabilities.

Figures 8a, b and c demonstrate the results of the Digital Capabilities barriers.

Regarding digital capacity, the results once again represent a strong barrier for companies, since the response rate between 4 and 5 is around 72%. According to [20], it is not enough to have managerial skills and specialized teams if there is a deficit in the ability to identify, acquire and implement new technologies.

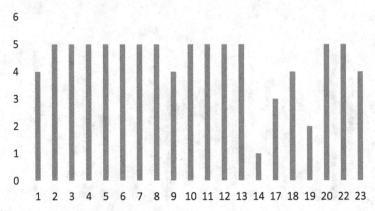

Fig. 8a. Barriers – We have the capability to identify digital technologies - "Digital Capabilities – Identification Capability"

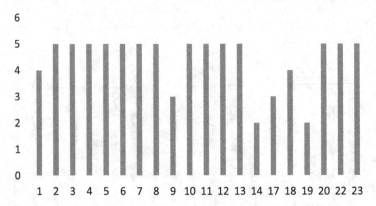

Fig. 8b. Barriers – We have the capability to acquire appropriate digital technologies - "Digital Capabilities – Acquisition Capability"

5   Global Discussion of Barrier Groups

In this section a global analysis of all barriers is performed: Big Data Analytical Capabilities, Dynamic Capabilities and Digital Capabilities. Figure 9 presents the results on the questions that had the highest response rates among the grades that express the greatest challenges in the Auditors' opinion. For better understanding and organization, barriers were named B1 to B28 (see Appendix A), where each of the horizontal axes represents an issue addressed in the questionnaire and analyzed in the previous section.

The results presented in Fig. 9 indicate that the biggest challenges in the implementation of an Analytical Big Data are concentrated between the B18 and the B28, which correspond to the issues relevant to Intangible barriers, Dynamic and Digital Capabilities, respectively. The most relevant are: managers' difficulty in coordinating activities related to the BDA in order to support other sectors of the company and external agents (B13); in the ability to absorb new knowledge that is relevant and generate value for

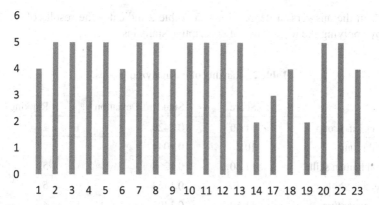

Fig. 8c. Barriers – We have the capability to implement digital technologies - "Digital Capabilities – Implementation Capability"

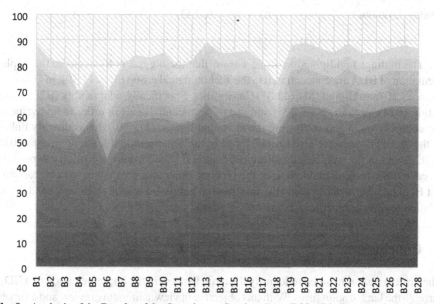

Fig. 9. Analysis of the Results of the Questions – Caption: B1 to B22 – Big Data Analytical Capabilities Questions; B23 to B25 – Dynamic Capabilities Issues; B26 to B28 – Digital Capabilities Issues.

the company (B20) combined with this, they also have great obstacles to acquire (B19) and apply (B21) this new knowledge; Finally, it is worth highlighting the challenges within digital capabilities that also permeate the ability to identify (B26) and acquire new technologies (B27). In addition, in Table 2 below, it is possible to observe, more broadly, the groups listed according to the degree of difficulty, taking into account the calculation of each of the answers obtained by applying a weight from 1 to 3, with 1 in the answers of degree 0 and 1, weight 2 in the answers of degrees 2 and 3 and finally, weight 3 in the answers of degree 4 and 5. Table 2 indicates the results of the Scores raised by applying the techniques of descriptive statistics.

Table 2. Ranking of the analyzed groups

Group	Score	Standard Deviation	Ranking
Technological group	11,80	0,1420	2
Financial group	10,2	0,1044	7
Technical human skills	11,40	0,1268	5
Management skills	11,40	0,1433	5
Data driven culture	11,50	0,1407	4
Organizational learning	11,52	0,1645	3
Dynamic capabilities	11,80	0,1853	2
Digital capabilities	11,87	0,2043	1

The findings highlighted in Table 1 show that the biggest bottlenecks in the implementation of BDA are concentrated in the Technological group (M = 11.80); in Dynamic (M = 11.80) and Digital (M = 11.87) Capabilities. Specifically, adaptability (B23), innovation (B24), organization (B25) and use of new technologies in order to add value to the company (B26 and B28) were the most highlighted (Fig. 9). This is directly linked to the fact that there are also serious flaws in the organizational learning groups (B18 to B22) and in the data-based culture (B16 and B17), since they are strictly dependent on each other for the functioning of the process, otherwise companies may be carrying out Big Data, but without using the information generated to add value and stand out in relation to their competitors, being only a waste of capital.

6 Conclusions

The present research aimed to identify the main barriers in the implementation of a BDA through the lens of audit firms. With this objective in view, a literature review and a survey were carried out through a questionnaire, which supported the infor-mation obtained in the analysis of results. The findings indicate that indicate that the main barriers to the implementation of a BDA in auditing companies are related to the dynamic and digital capabilities of the companies, that is, their ability to adapt to new chang-es, both in the market and in technology, as well as their ability to identify and im-plement new technologies in their organizational routine. In addition, it is possible to observe that the issues involving the organizational learning of the company and the data-based culture are compromised, since the dynamism and technology are not well structured. A company cannot have a data-based decision-making culture if they do not have a structure for this data to be credible enough, thus perpetuating decision-making based on subjective judgments by managers.

However, even with these challenges highlighted, it can also be concluded that there are other issues, such as the financing of the tool, the search for qualified employees and the training of the existing team, which are being addressed more easily by com-panies even if they have limitations. to be overcome. In summary, this evolution is enough for us to have an optimistic perspective on the evolution of Analytical Big Data within the audit sector, so that it will add value and even transform the way the process is currently carried out. Given the above, it is essential that managers focus their efforts on the continuous improvement of their dynamic and digital capabilities in order to reduce the distance from the implementation of such an important tool and heavily used in other sectors of different industries, bringing a new barrier to competitiveness. In addition, it is also up to managers to continue with the initiatives to train teams and inject funding into Big Data projects, since they are equally important for their operation.

The authors of this work recognize that this study has some limitations that must be addressed. In our research structure, another limitation is related to the sample size of this study. Enlarging the sample would be interesting in future research. This study was applied in the context of Brazil. Thus, developing the research in other countries is recommended for future studies in order to compare which barriers in the implemen-tation of big data analytical capacity are more dominant. Another limitation concerns the statistical method used, which was based on descriptive statistics. Expanding data processing methods (Thurstone's Law of Categorical Judgments, Artificial Neural Net-works, etc.) is recommended to analyze audit firm questionnaire responses to provide more insights into BDA in audit firms.

Since this is a new and little explored subject, this study adds to the literature new information about what are the main barriers faced in practice by companies, leaving the theoretical field alone. This study also contributes to the practice of Auditing in Brazil and the big data analytics represents a field of opportunity [16]. It is hoped that this research will serve as a basis for future research that can be guided always aiming at improvement with regard to the barriers of a BDA. This study has limitations, specifically about the sample. Thus, it is recommended to expand the sample for future studies. Despite the limitations, the research is relevant and generates insights for future studies in the field of Auditing.

7   APPENDIX A – Barriers to Implementing BDA

Big Data Analytical Capabilities	TANGIBLE	Data	We have the ability to access large volumes of data	B1
			We have the ability to integrate multiple internal data sources for analysis	B2
			We are able to integrate multiple external data sources for analysis	B3

(continued)

(*continued*)

		Basic features	We have the ability to invest financial resources in analytical Big Data projects	B4
HUMAN SKILLS		Technical abilities	We offer training oriented to big data analysis for professionals/employees	B5
			We always hire new professionals with experience in big data analysis	B6
			Our big data analytics team has the right skills to do their jobs successfully	B7
			Our big data analysis team is well trained to perform their duties	B8
			Our big data analytics team has adequate work experience to do their jobs successfully	B9
			Our big data analytics team is well trained	B10
		Management skills	Our big data analytics managers understand and appreciate the business needs of other functional managers, suppliers and customers	B11
			Our big data analytics managers can work with functional managers, suppliers and customers to determine the opportunities big data can bring to our business	B12
			Our big data analytics managers are able to coordinate big data related activities to support other functional managers, suppliers and customers	B13
			Our big data analytics managers are able to anticipate the future business needs of functional managers, suppliers and customers	B14
			Our big data analytics managers have a good sense of where to apply big data	B15
INTANGIBLE		Data Driven Culture	We have a data-driven culture	B16
			Our decisions are based on data	B17

(*continued*)

(continued)

		Learning	Do we share knowledge internally and externally with partners – stakeholders?	B18
			We have the ability to acquire new and relevant knowledge	B19
			We have the ability to absorb relevant knowledge	B20
			We have the ability to apply relevant knowledge	B21
			We have the ability to explore new knowledge	B22
Dynamic Capabilities			We are capable of adapting/reconfiguring to changes	B23
			We have the capacity for technological innovation	B24
			We have organizational capacity	B25
Digital Capabilities			We have the ability to identify digital technologies	B26
			We have the ability to acquire appropriate digital technologies	B27
			We have the capacity to implement digital technologies	B28

(Source: Adaptation from Mikalef et.al., 2019).

References

1. Marshall, A., Stefan, M., Shockley, R.: How leading organizations use big data and analytics to innovate. Strategy Leadersh. **43**(5) (2015)
2. Mikalef, P., Boura, M., Lekakos, G., Krogstie, J.: Big data analytics and firm performance: findings from a mixed-method approach. J. Bus. Res. **98**, 261-276 (2019)
3. Cressey, D.R.: Other People's Money: Study in the Social Psychology of Embezzlement. 1st edn. Wadsworth Publishing Company, Editora (1973)
4. Raj, P., Deka, G.C.: A deep dive into NoSQL databases: the use cases and applications. Advances in Computers, Book series, vol. 109, pp. 1–390 (2018)
5. Tang, J., Karim, K.: Financial fraud detection and big data analytics – implications on auditors' use of fraud brainstorming session. Manag. Auditing J. **34**(3), 324–337 (2018). https://doi.org/10.1108/MAJ-01-2018-1767/full/html. Accessed 08 Feb 2022
6. Alles, M., Gray, G.L.: Incorporating big data in audits: identifying inhibitors and a research agenda to address those inhibitors. Int. J. Account. Inf. Syst. **22**, 44–59 (2020)
7. Shukla, M., Mattar, L.: Next generation smart sustainable auditing systems using big data analytics: understanding the interaction of critical barriers. Comput. Ind. Eng. **128**, 1015–1026 (2019)

8. Reurik, A.: Financial Fraud: A Literature Review. Max Planck Institute for the Study of Societies, Cologne (2018) https://doi.org/10.1111/joes.12294. Accessed 8 Feb 2022
9. Young, C.: Financial Statement Restatements: Trends, Market Impacts, Regulatory Responses, and Remaining Challenges. U.S. Government Accountability Office (2002). https://www.gao.gov/products/gao-03-138. Accessed 08 Feb 2022
10. Buhl, H.U., Roglinger, M., Moser, D.K.F. Heidemann, J.: Big data: a fashionable topic without sustainable relevance for research and practice?. Bus. Inf. Syst. Eng. (2013). https://aisel.ais net.org/bise/vol5/iss2/1/. Accessed 16 Jul 2022
11. Zhong, R.Y., Huang, G.Q., Lan, S., Dai, Q.Y., Chen, X., Zhang, T.: A big data approach for logistics trajectory discovery from RFID-enabled production data. Int. J. Prod. Econ. **165**, 260–272 (2015)
12. Keeso, A.: Big data and environmental sustainability: a conversation starter. Smith School Working Paper Series (2014) http://www.smithschool.ox.ac.uk/publications/wpapers/workin gpaper14-04.pdf. Accessed 16 Jul 2022
13. Jharkharia, S., Shankar, R.: IT-enablement of supply chains: understanding the barriers. J. Enterp. Inf. Manag. **18**(1), 11–27 (2005)
14. Luken, R., Van Rompaey, F.: Drivers for and barriers to environmentally sound technology adoption by manufacturing plants in nine developing countries. J. Cleaner Prod. (2008). https://www-sciencedirect.ez24.periodicos.capes.gov.br/science/article/pii/S09596 52607002090 Accessed 16 Jul 2022
15. Rakipi, R., De Santis, F., D'Onza, G.: Correlates of the internal audit function's use of data analytics in the big data era: global evidence. J. Int. Acc. Auditing Taxation **42** (2021)
16. Earley, C.E.: Data analytics in auditing: opportunities and challenges, **58**(5), 493–500 (2015)
17. Gepp, A., Linnenluecke, M.K., O'Neill, T.J., Smith, T.: Big data techniques in auditing research and practice: Current trends and future opportunities. J. Account. Lit. **40**, 102–115 (2018)
18. Cao; M., Chychyla, R, Stewart, T.: Big data analytics in financial statement audits. Account. Horiz. **29**(2), 423–429 (2015)
19. Dagilienė, L., Klovienė, L.: Motivation to use big data and big data analytics in external auditing. Manag. Audit. J. **34**(7), 750–782 (2019)
20. Gupta, M., George, J.F.: Toward the development of a big data analytics capability. Inf. Manag. **53**(8), 1049–1064 (2016)
21. Staisey, N.: Big Data Raises Big Questions. Government Technology (2013). https://www.govtech.com/archive/big-data-raises-big-questions.html. Accessed 16 Jul 2022
22. Ross, Jeanne, W., et al.: You May Not Need Big Data After All. Harvard Business Review (2013). https://hbr.org/2013/12/you-may-not-need-big-data-after-all. Accessed 16 Jul 2022
23. Bhatt, G.D., Groover, V.: Types of information technology capabilities and their role in competitive advantage: an empirical study. Taylor and Francis Online **8**(2), 253–277 (2014)
24. Davenport, T.: Big Data at Work: Dispelling the Myths, Uncovering the Opportunities. Harvard Business Review Press, Editora (2014)

Applicability of Fractal Architecture Based Microservices on System-of-Systems

Bala Krishna Dhamodaran(✉) iD

Department of Artificial Intelligence and Systems Engineering, Riga Technical University, Riga 1048, Latvia
bala-krishnaDhamodaran@edu.rtu.lv

Abstract. Microservices is a set of independent software developments communicating over APIs (application program interfaces) to achieve a whole generic system. This paper aims at addressing the usefulness of a fractal architectural design in modern organization with an IT (Information Technology) orientation and a variety of businesses under single entity, on structuring and integrating such microservices on system of systems. It is crucial to maintain proper alignment concerns and autonomy concerns while adhering to the general purpose of the business and considering other factors such as competition. Fractal architecture provides a conceptual framework for a balanced trade-off between autonomy and alignment concerns. This study analyses the feasibility of fractal architecture, its opportunities and challenges based on previous literature when applying to microservices on system of systems. For this purpose, various metrics defined by the authors are used to achieve research objectives and identify limitations of fractal architecture.

Keywords: Fractal Architecture · Microservices · System of Systems

1 Introduction

With the increasing sophistication and dynamics of enterprise environment, the need for identification of a robust and scalable architectural design for enterprises is vital. Due to increasing competition and dynamic customer needs, in a Cyber-Physical environment to optimize the profit and reduce cost enterprises should look beyond traditional existing monolithic approach which is designing single executable artifacts [1]. To address the modern scenario, a more cohesive and independent approach is to be initiated. This is where microservice architecture comes into play. It is an architectural model inspired by service-oriented computing [2]. In microservice architecture components of the system can be developed and deployed independently, thus addressing modern scalability issues imposed by external factors. Scalability is a major concern in the context of system-of-systems (SoS), which is a collection of operationally and managerially independent set of individual systems working together to achieve a common set of goals collectively [3]. Depending on the objective, governance, and interrelationship of the constituent systems, [4] have classified SoS into four main categories namely directed,

J. Maślankowski et al. (Eds.): PLAIS EuroSymposium 2022, LNBIP 465, pp. 109–125, 2022.
https://doi.org/10.1007/978-3-031-23012-7_7

collaborative, virtual, and acknowledged. In the directed type, the system is centrally managed and focuses only on a specific purpose. While in the collaborative type, various components would collaborate to run the system without a well-defined power vested on the central management. Moreover, virtual SoS would function on the basis of the interaction between various components without a central management and a defined purpose. Lastly, acknowledged SoS is a combination of both directed and collaborative SoS, where there is a central management; however, the components would have their full independence and are responsible for any changes in SoS. These system-of-systems are enabled by novel technologies and concepts such as smart cities, smart hospitals, and remotely operated factories.

For the proper manifestation of microservice architecture on system-of-systems, synergy between these concepts is to be achieved [5]. To manage system-of-systems that aims at achieving a common goal, self-similarity of different constituents is needed at granular levels which are required by SoS in achieving a common goal. This feature also helps system operators as well as system maintenance departments to diagnose any malfunction in the system effectively. Moreover, self-organization and self-optimization are required to maintain the independence of various components at various levels. These features are achieved by a fractal architecture [6], where it is a conceptual framework centered around a technical integration pattern which is built recursively. Fractal framework is aimed at facilitating homogeneity in granular level in the same fractal level of operation while maintaining autonomy in the whole system. Microservices based system-of-systems is a concept where microservices are implemented in constituent components of the whole system to achieve granular level goals and thus, addressing a common goal. This paper focuses on applicability of a fractal framework which facilitates microservices based system-of-systems by a systematic literature review.

For this research following research questions are formulated and addressed:

RQ1: What are the driving factors behind the use of fractal architecture for microservices-based system-of-systems?
RQ2: What are the benefits of utilizing fractal architecture on microservice based system-of-systems?
RQ3: What are the constraints of deploying fractal architecture into existing and future system-of-systems?

The rest of the paper is organized as follows. Section 2 discusses the background of microservices based system-of-systems and fractal architecture in detail which are available in previous literature. Section 3 presents the research procedure alongside with exclusion and inclusion criteria. Section 4 discusses the findings and results while identifying the features of fractal architecture which make it an applicant for microservice based system-of-systems, its benefits, and limitations. Finally, Sect. 5 concludes the research having identified future research directions under the topic of interest.

2 Background

This section briefly describes the already implemented fractal-based architectures, while introducing fundamental concepts pertaining to microservice architecture and system-of-systems. Additionally, a generic system-of-systems model based on microservices with fractal architecture is presented.

2.1 Fractal Architecture

Fractal architecture is based on the "fractal company" concept presented in [7] where the basic emphasis is on similarity at various levels of granularity of the enterprise, which are acting independently towards a concise goal. The basic unit comprising this system is known as a fractal. As mentioned by Warnecke [7], the main properties of fractal-based systems are *self-similarity, self-organization, goal orientation, dynamics,* and *vitality.* Self-similarity comes with the fact that fractals can act as a black box, where exact output is obtained when the same input is given to fractals regardless of their internal structure. Fractals practice self-organization both operatively and strategically. They are able to achieve their goals in an optimal organization by applying appropriate methods; for this purpose, they will restructure, regenerate and evolve. Goal-orientation is due to the fact that fractals are capable of negotiating with other fractals to attain their goals and they are also capable of formulating and modifying goals, as necessary. Since fractals are networked to maintain cooperation and coordination between other fractals, they can adapt to dynamically vibrant environments.

As presented in [8], the fractal model can be applied at two levels of abstraction, one at the level of business process system and another of software system. At the business level, fractal model can be used for facilitating business process flexibility, while at the software system level, it enables flexibility on software design that supports the business processes. Moreover, Stecjuka et al. [8], have identified two approaches in implementing a fractal architecture. In a process-oriented approach, the fractal properties at the business process level are identified first and then the software requirements are structured in accordance with these processes. In contrast, in the object-oriented approach, fractals corresponding to the software system levels are identified by defining goals, and the formulated fractal software architecture is then introduced for business system support. Corresponding steps for each approach are addressed in [8].

2.2 Microservice Architecture

Microservices architecture (MSA) is a modern trending paradigm used by many of the leading digital enterprises such as Netflix, Spotify, Amazon which are functioning on a diverse and dynamic environment [9]. The main driving force behind the migration from more monolithic architectures to microservice architecture is the ease of scalability embedded with microservice approach alongside simple communication protocols when compared to service-oriented architecture (SOA) [10]. However, both SOA and microservices architecture proposes decomposition of systems into services which are available over a network, but in SOA the services are connected via smart routing mechanism with a central governance where, in contrast, microservices architecture lacks a central governance and employs simple routing mechanisms. Taibi et al., [9] have presented a pattern catalogue for differentiating various microservices architecture based on previous literature. They have categorized these architectures broadly under three categories: patterns pertaining to different orchestration and coordination-oriented architectures, patterns reflecting various physical deployment strategies and patterns utilizing different data storage patterns. Moreover, microservices architecture offers advantages such as scalability, reliability, ease of deployment, technology heterogeneity, organizational alignment [1, 11] which makes microservices architecture an exceptionally suitable candidate in smaller well-partitioned web-based applications [10], IoT (Internet of Things) applications [11], cloud-based services such as Software-as-a-Service [2] and more complex systems such as smart city concepts [12] and smart grids [13]. However, adaption of microservices architecture for system scalability is to be analyzed which would otherwise hinder goal accomplishment required by the system. A proper systematic scalability analysis along with a goal-obstacle analysis is carried out by [14] by utilizing keep-all-objectives-satisfied (KAOS) goal-oriented modelling.

However, as highlighted in [1], there are trust and security issues in microservices implementation: greater surface attack area due to open API (Application Programming Interface) communication over the network, complex network activity which would cause limitations in implementing security measures, heterogeneity can also cause complexity, and lack of central governance can impose trust issues. Furthermore, Carrasco et al. [15] have identified nine different pitfalls that are encountered in migration towards microservices under two broad categories of architectural and migration pitfalls [15]. However, [16] have successfully applied an approach based on the concept of bounded contexts for a microservice decomposition of a large industrial Enterprise Resource Planning software. Also, various research has been carried on trends and potential of microservices for industrial adoption, [17] is a notable reference, and attempts to make autonomous microservice architectures are also prominent [18].

2.3   Features of System-of-Systems

System-of-systems, which comprises of independent constituent systems working towards a specific goal. As identified in [5], broad six criteria have been proposed to clearly identify a system-of-systems. They are:

1. Operational independence: Every constituent system should operate independently to fulfil specific set of goals.
2. Managerial independence: Every constituent system which is assembled as part of SoS should operate under any circumstances on its own.
3. Geographical distributions: Constituent systems are distributed over range of geographical location(s).
4. Emergent behavior: The overall behavior of a SoS is a collaborative behavioral result of constituent systems, which does not depend on isolated behavioral pattern of each system.
5. Evolutionary development: A SoS continuously evolves by modifying its structural and functional aspects.

6. Heterogeneity of constituent systems: The comprising constituent systems have different natures that would lead them to operate on different time scales and dynamics.

In addition, [5] has discussed the use of similar traits in a microservice architecture (MSA). It has also discussed the importance of microservice-based systems as a more well-designed approach toward realizing an SoS. The major contribution of the work was to identify the defining features of MSA that can be realized as the main facilitating features of SoS.

Componentization via services is an important feature in which the defining components of an MSA are implemented as units of software developed as *services*. This feature helps to achieve managerial and operational independence in a system and eases in evolutionary development. Moreover, the componentization promotes deployment of the system on a vast geographical scale while enabling heterogeneity of constituent systems.

MSA is organized around business capabilities, where a system is developed under full-stack development architecture while focusing on a pre-defined business line and objectives. This enables operational and managerial independence and assist in distributing the system depending geographical location. The need for a distributed geographical location may be due to the magnitude of business demands and processes.

An interesting feature that is strongly related to organization around business capabilities is decentralized governance and data management. This largely eases operational and managerial independence as each cross-functional team developing different functions of the system can function independently. Additionally, with the rise of independence in the system, the emergent behavior of the overall system becomes more prominent. It also leads to the ease of distribution of the system over a wide geographical profile. Finally, this ensures evolutionary development of the system, as data and governance aspects can evolve independently.

Furthermore, since MSA is formulated while focusing on scalability and replaceability, with fast responses to system failures, it enables evolutionary development of the system. Automated deployable infrastructure enables operational and managerial independence and let the designers focus more on the overall system alongside expected business capabilities. These features of MSA can enable well-designed SoS architectures.

2.4 Facilitating Microservices Architecture on System-of-Systems

Microservices architectures plays a significant role in enabling system-of-systems. In this regard, [5] has discussed features of microservices facilitating proper synergy between system-of-systems. Since microservices are broken down into components, which are deployed as software units, they can achieve several features like that of SoS. Microservice architecture, is based on "smart endpoints and dumb pipes" concept, here the endpoints or the deployed software units are smart and capable of evolving while they are communicating with each other using simple protocols such as HTTP. Moreover, decentralized governance and addressing business capabilities at granular level makes microservice architecture a catalyst in enabling SoS features.

A generic system-of-systems model based on microservices with fractal architecture is shown in Fig. 1. In the left-most image, a SoS is shown which consists of constituent systems, these constituent systems are made from fractals at different granular levels. These constituent systems communicate with each other using communication links. Moreover, as shown in the right-most image, the comprising fractal layers have microservices running on each layer to achieve autonomy along the vertical axis while maintaining homogeneity of microservices along the horizontal axis.

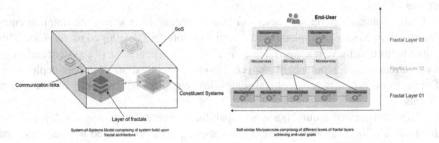

Fig. 1. System-of-systems model based on microservices with fractal architecture

3   Research Method

To realize the applicability of fractal architecture on microservice based system-of-systems a systematic literature review process was carried out according to Kitchenham and Charters report [19].

In this process, the first step is to identify the need for research, this phase establishes the domain for future research. Next, research questions are formulated, which will help to narrow down the research and identify the scope of the research. Having identified research framework and scope, a suitable search strategy is to be used to find relevant literature. The deployed search strategy should be complete and sufficient to gather information to address the formulated research questions. After gathering relevant literature, relevant data should be extracted from the literature to do a comprehensive analysis of the data. As the final steps, a discussion is presented along with results from the data extracted and analysis addressing the formulated research questions and then the study is concluded alongside with any future research directions.

3.1   Search Strategy

This subsection describes the employed search strategy used for selection of relevant literature. Initially, a list of keywords is identified, these were identified using research questions, a preliminary review of literature using web search.

Identified Keywords: fractal-model, fractals, microservices, microservices architecture, fractal IT, decomposition of microservices, system-of-systems, fractal architecture, business modeling, fractal company.

Secondly, these keywords were used for development of search string. The search string was used for searching articles. The search string was checked on article title, abstract and keywords. Phrases comprising more than one word were mentioned within "" to obtain the exact word matching.

Search String: Microservices AND (fractal OR "fractal model" OR "fractal architecture") AND (systems OR "system-of-systems" OR enterprises).

Thirdly, the search string was used in the following libraries: Scopus, ACM Digital Library and Science Direct. Also, Google Scholar search engine was used. Moreover, snowballing technique was used to investigate further literature.

Lastly, the exclusion and inclusion criteria of obtained literature were stated:

C1: Literature should be comprised of journal articles, conference proceedings, books.
C2: Selected literature should be in English language. Articles from other languages were excluded with articles from year 2006 onwards.
C3: Articles that were not relevant to the topic and subject domain were excluded.
C4: Full access to text should be available.

The literature search strategy process in the selected databases and search engine and the criteria applied to those studies are also included. Initially, 209 articles were selected using the search string on databases and Google Scholar. Then, second criteria of selecting articles only in English language and articles from year 2006 onwards were filtered. This result in 199 articles. Then, articles related subject domain were selected this result in 38 articles. This was carried out by scrutinizing the abstract. Lastly, full-text articles with public access were selected. This results in 31 articles in total. The process is shown in Fig. 2.

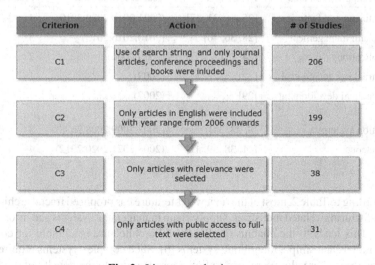

Fig. 2. Literature selection process

4   Results and Findings

This section will present various findings concluded and extracted from literature and will address the formulated research questions. The reviewed literature spans in three areas of interest, which is listed in Table 1, alongside with their frequency of appearing in final 31 articles.

Table 1. Reviewed literature based on subject domain

Subject domain	References	Frequency
Microservices related	[5, 9–12, 14–17, 20–22]	12
Fractal model based	[8, 21, 23–40]	20
System-of-systems related	[5]	1

In Table 1, the subject domain is cross checked with the title, keywords, and abstract of the paper. Next, the focus was given to identifying various application areas of fractal architecture in literature. For this purpose, a set of industries and application areas are identified, and the reviewed literature is sorted under each category which are presented in Table 2.

Table 2. References with proposed or implemented fractal architecture

Industry / Application area	References	Year of Publication	Frequency
Manufacturing and factories	[25, 26, 31, 37]	(2021,2010,2020,2019)	4
Supply chain management	[28, 36, 40]	(2010,2015,2013)	3
Web applications	[24, 27]	(2019,2006)	2
Quality tracking and control	[32]	(2008)	1
Pharmaceutical development process	[29]	(2007)	1
Information systems	[8, 23, 34]	(2008,2009,2008,2008)	3
Miscellaneous	[21, 38, 39, 41]	(2008,2021,2020,2012)	4

According to Table 2, most of the reviewed literature has proposed fractal architecture for manufacturing-related industries, supply chain management systems and information systems. This leads to the conclusion that fractal architecture is suitable for complex systems rather than simple monolithic systems [6]. Such complex systems require more sophistication and are closely related to SoS. However, due to the novelty of the fractal architectural design in microservice domain, more research needs to be carried out to rationalize the proposed approach. More detailed analysis of reviewed literature is provided below.

- Visotschnig [25] has proposed a fractal control system architecture for next-generation factories. In the presented conceptual system, process automation is achieved through cyber-physical fractal nodes. These fractal nodes are called Fractal Node System (FRANS) and are categorized into seven categories based on their functionality. These are measurement, action, preprocessing and analysis, optimization, data management, visualization, control, and surveillance nodes. These FRANS nodes are implemented as container image which are self-contained and transportable. The major limitation identified in the work is the limited performance of the underlying communications system and computing resources hosting FRANS nodes. Future work focuses on implementing the proposed system in industrial use cases for further optimization and evaluation.

- Ryu [26], focuses on an agent-based fractal architecture for distributed manufacturing systems. The study has identified basic fractal unit (BFU) as the major component, with five functional modules: observing module (observer), analyzing module (analyzer), resolving and executing module (resolver), organizing module (organizer), and reporting module (reporter). However, the study declares that the implementation of fractal architecture is difficult despite its advantages. In order to develop these models, the Unified Modelling Language (UML) is used, as it can be modelled from various points of view. Since, it was UML models impose limitations in forecasting and prediction, Petri-nets were used to predict the future status of the system. The work is limited by a conceptual development and aims at developing a prototype using the Object Constraint Language (OCL).

- [31] has evaluated the feasibility of using the fractal architecture to develop cyber physical systems (CPS) in Industry 4.0. For this purpose, the authors have evaluated the previous literature and identified two main challenges in integrating fractal architecture into CPS: (1) virtualization using CPS and (2) sustainability requirements. In the context of realizing sustainability requirements, the main drawback of fractal theory is that it does not incorporate any environmental criteria. [37], on the other hand, proposes reference architecture by abstracting functions and structures of a CPS system in terms of resource fractals (RF). The effectiveness of the system was verified by a case study of the manufacturing process of a cylinder product. The case provided insight on how the fractal architecture eases the business processes as well as the production processes.

- Study [28], has identified use of fractal-based supply chain management of e-Biz companies. The type of company focused in this paper is a B2C (business-to-customer) which connects customers with product manufacturers via a web interface. This study identifies five functional units: observer, analyzer, resolver, organizer, and reporter. These functions are modeled using UML. In addition, the study also presents a profit model for the proposed e-Biz company with a numerical example. However, the paper has limitations in the proposed mathematical model as the objective function of the model only considers delivery cost and doesn't include costs such as trust in transportation and customer satisfaction.

- More research on supply chain management base on fractal theory is carried on [36], where the aspect of concept of fractal supply chain is analyzed based on its characteristics, structure and functional model. It has also developed a detailed supply chain design procedure. The basic steps in the formulated design procedure are: analyzing the industry environment, structural adjustments to facilitate self-optimization and self-organization and finally operation and improvement. In contrast, [40] have focused on utilizing a fractal-based echelon to minimize inventory costs and maintain a smooth material flow. [40] also has carried out a comprehensive simulation model to compare traditional vendor managed inventory (VMI) and fractal-based VMI (fVMI). According to the results of the study, the cost savings from fVMI are significantly higher than that of traditional VMI for all levels of demand variability.

- Koleini *et al.* [24], have presented a fractal application capable of automated scaling on its own. This can be seen as an alternative approach to cloud service management which allows embedding of service management in the application logic of the service. Services are controlled by simple Remote Procedure Call (RPC) APIs on the Jitsu control stack and allow simple, dynamic, programmable creation and destruction of services. The network-based load balance is ensured by the OpenFlow protocol. As a proof of concept of the proposed architecture, they have developed a self-scaling web service on a MirageOS webserver. The proposed system effectively provides a variable load distribution between replicas, while having no substantial negative performance impact.

- A similar study [27] has carried out research on the Fractal Component Model and its support in Java. Components of this proposed model can be assigned with different capabilities. The study describes a small runtime environment known as JULIA, a Java implementation of the model. This is used to program the reflective features of the components of the model. The effectiveness of the model was evaluated by reengineering an existing open-source message-oriented middleware. The results showed that performance of JULIA, is highly favorable over standard Java implementation functionality.

- [23] have conducted extensive research on realizing and steps in designing and tuning information systems (IS) to incorporate fractal architecture. In order to depict the practical applicability of the fractal paradigm in developing IS, announcement services for the university were developed in related research. The study proposes further research on integrated modelling of all components in IS in terms of fractal entities and changes in the propagation networks involved. In contrast, related work [8] evaluates the applicability of the fractal paradigm at two levels of abstraction: the level of the business process system and the level of the software system. It also identifies a limitation of the fractal paradigm where it can be applied only to business process and software subsystems that exhibit fractal properties in at least one modeling dimension.

- [34] has carried out research on development of fractal-based Information System IS, platform independently by using Model Driven Architecture (MDA). The main objective of employing such design architecture is to enable portable software designs and to increase the reusability of the designs. Applying the proposed method for enterprise applications, remain as future research.

- Xu *et. al.* [32] have studied the use of fractal and mobile agents to tackle the complicated problem of quality tracking and control between enterprises in networked manufacturing. For this purpose, they have proposed the Inter-Enterprise Quality Tracking and Control (IeQTC) framework. The proposed framework has advantages particularly in distributed and dynamic environments, due to inherent features of fractal architecture such as self-similarity and self-organization. The flow model of the fractal process is realized using fractal mobile agent (FMA) templates. By using these templates, distributed and mutually related information is collected, transferred, processed, and tracked.

- [29] presents a conceptual model for the networking of small and medium enterprises (SMEs) based on fractal architecture. The developed SME networking model is justified with a case study related to the process development of a new pharmaceutical product in an SME network.

- Deployware [21] is a distributed and heterogenous software system deployed on grids based on fractal architecture. Another interesting research is carried out in [38] to a Fractal Enterprise Model (FEM) on case study based on a water utility company. They have also identified problems that were solved from deploying a FEM on the system. These include understanding the domain and the business context and business operation and defining changes in the business and its future scope. However, the study also has identified three main limitations in deploying FEM. Firstly, such system has only one relation to indicate that an asset is a software system or a data entity. Secondly, FEM did not provide means for depicting detailed characteristics of assets and processes. Finally, using FEM needs some prior training.

- Another application of fractal architecture in information systems is discussed in [39], where fractal features are incorporated into the systems of higher education institutes in Kurdistan. The study has studied previous literature and has successfully justified the use of a fractal architecture in education information systems to attain independent units in a distributed system. [41] has compared a traditional environmental management system (EMS) with a fractal EMS (FEMS).

After reviewing the literature, various fractal approaches are identified. These are shortly presented in [8] and summarized into two broad categories as process-oriented and object-oriented. Figure 3 shows the usage of various fractal approaches used in the literature. [8] has used a hybrid approach of both service and object approaches to address the level of the business process and then to structure software requirements and define the software architecture for business support, respectively. The reviewed literature composed of hybrid approaches that slightly deviate from these approaches, [20] has suggested a rank-based fractal architecture for inventory management in supply chain networks. Moreover, the concept of multi-fractal approach is employed where many fractals function together. They maintain self-similarity among the components and work towards a defined goal. This approach is used in systems where there is a need for scale invariance ensuring the evolutionary design paradigm of the system.

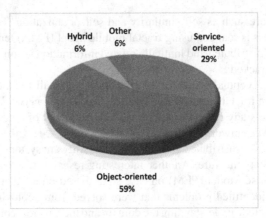

Fig. 3. Fractal approaches used by literature

The following subsections will address the formulated research questions and how they are addressed using the results and findings.

4.1 Factors Facilitating Use of Fractal Architecture on Microservices Based System-of-Systems

In the context of a generic system, an external and an internal environment can be identified. A proper balance of factors affecting under these environments to the system as a whole is important. First, it is important to distinguish various features in both internal and external environments. With the advancement of complex and dynamic user demands, scalability of systems to address competition and flexibility of vigorous Cyber-Physical domain in which modern systems operate are the characterizing features of the external environment. Key features related to the internal environment are the heterogeneity of various levels of systems that operate at different granular levels and a strong contingent language barrier between developers and system architects. To address these issues a self-evolving and self-organizational architecture is required. This is where existing Mutually Exclusive, Collectively Exhaustive (MECE) approaches fail [6]. As MECE approaches focus on more standardized classical management and emphases on avoiding redundancy. However, in order to realize a self-evolving and self-organizational architecture, redundancy is an essential requirement.

Maintaining a balanced trade-off between autonomy and alignment concerns is vital in all levels of system granularity. Fractal architecture possess autonomy capability where each microservice relevant to a given function has the independence of carrying out a specific task and controlling the relevant resources. A major alignment concern is also maintained by implementing fractal architecture on microservices as they are goal-oriented and would attain their goal either by negotiating with other fractals or optimizing their resource usage as seen in [21]. Self-similarity and self-organization are other key features that assist to address heterogeneity of systems that also hinder scalability of systems in dynamic environment. This also provide an opportunity to develop the system in a recursive manner which enhances learning and self-optimization of microservices in a system.

4.2 Benefits of Fractal Architecture

By incorporating fractal framework onto modern enterprises that are operating in Cyber-Physical domain will have an opportunity in providing an integrated user experience. This is facilitated by the ability of maintaining homogeneity of microservices at different granular levels of a given fractal framework. The end-user will not have to worry about the underlying microservices in an enterprise, despite differences in services they provide they will achieve the common goal collectively as discussed in the proposed in study [28], where an e-Biz company based on supply chain management is evaluated.

Microservices, which are operated in cloud would require scalable requirements as integration of new customer demands regarding new products or optimization of services. These can be easily addressed by utilizing a fractal framework which would provide self-optimization and dynamic nature to the microservices to evolve and overcome the issues successfully. Another, important aspect of enabling the fractal framework is the increase of traceability of business processes. This is enabled by the self-organizational and self-similarity of fractals. This is aided by lightweight event format [6] incorporated into fractal architecture.

Moreover, as discussed in [8], fractal modeling approach can be extensively used for business process flexibility. Providing flexibility to business processes can be optimized to either minimize costs or maximize profit of business processes.

4.3 Limitations of Fractal Architecture

As presented in [6], since fractal architecture is applied to larger landscapes of enterprises, an inherent complexity in both implementation and design is observed. Naturally, making fractal approach less competent for small IT-based business processes. Therefore, need for sufficient complexity in a system, is a limitation when implementing fractal architecture. Also, since in fractal architecture software services or microservices are implemented as deployment units, design of a cohesive or a single entity of software for a process is improbable. Furthermore, since fractal architecture does not dive into details of cyber-physical infrastructure on which business processes and services are carried on, it is less self-aware about the infrastructure capabilities. This is another limitation in fractal architecture, as the system can be limited by its infrastructure capability. Lastly, lack of central governance can provoke issues when it comes to controlling and maintenance where no credibility is assured.

5 Conclusion

This research is a brief survey of various fractal architecture based microservice implementations in system-of-systems. The systematic review, enabled to envision the applicability of fractal architecture on such systems, due to its attractive features such as scalability and flexibility in the context of modern Cyber-Physical systems. From the review of articles, it can be concluded that fractal architecture is more applicable for more complex system-of-systems as the application area of "Manufacturing" had the highest number of frequency of articles under the topic of interest. Apart from this application area, "Information Systems" and "Supply Chain Management" had next highest incorporation of fractal architecture into system design. Further, all these application areas are modelled using various microservices running on different constituent systems. Some studies showed an interest on decomposition of these applications to achieve more robust and modern systems, while some studies presented metrics in determining the proper decomposition of microservices. However, since the use of fractal architecture is novel, additional research is to be carried out as future work and study results should be validated against a much larger sample of studies.

The formulated research questions were addressed where detailed findings were presented. Initially, the factors facilitating fractal architecture on microservice based system-of-systems were evaluated. One of the main facilitating factors is the need for self-evolving and self-organizational architecture. Equally, fractal architecture is important in enabling a balanced trade-off between autonomy and alignment concerns in a system. Finally, it is valuable in systems where self-similarity and self-optimization are important to enable evolutionary design concepts and to maintain operational and managerial independence. Benefits of enabling fractal features include provision of an integrated user experience, fulfilling scalability and reusability requirements in a dynamic environment, and increased traceability of business processes. Moreover, fractal architecture enables flexible design in business processes.

Research also identified limitations of fractal architecture enabled systems. Lack of central governance can lead to issues that arise when it comes to controlling and maintaining the system. Due to the natural complexity of fractal architecture, there can be complexities in deploying a fractal architecture to a simpler system, as it can further increase the complexity of the system. Finally, the fractal architecture can exploit the available infrastructure facilities, as it is less aware of infrastructure capabilities and more concerned with business goals and capabilities.

Proposed future work is to readily implement fractal architecture on industrial level use cases. Also, apart from microservices, a much concern can be given to serverless functions or Functions as a Service (FaaS) to act as fine-grained units in modern enterprises and possibility of applying fractal architecture onto them [6]. According to Conway's law, the structure of the built-in software will mirror the structure of the organization. In this context, a future research direction would be to understand how fractal architectural designs can incorporate end-user communication into its design paradigm. Furthermore, integration of different fractal approaches, i.e., process and object-oriented approaches to achieve process flexibility simultaneously from both business and software perspectives remain yet another research direction. Although Service Oriented Architecture (SOA) is becoming obsolete, possibility of return of such advances would be possible, hence the applicability of fractal architecture in such contexts is also an interesting research direction identified through this research [6].

References

1. Dragoni, N., et al.: Microservices: Yesterday, today, and tomorrow. Present Ulterior Softw. Eng., pp. 195–216, November 2017. https://doi.org/10.1007/978-3-319-67425-4_12. Accessed 13 Jun. 2022
2. Familiar, B.: Microservices, IoT and Azure: Leveraging DevOps and Microservice Architecture to Deliver SaaS Solutions. Apress (2015)
3. Assaad, M.A., Talj, R., Charara, A.: A view on Systems of Systems (SoS). In: 20th World Congress of the International Federation of Automatic Control, no. 1991 (2016)
4. Collins, B., Doskey, S., Moreland, J.: Relative Comparison of the Rate of Convergence of Collaborative Systems of Systems: A Quantified Case Study
5. Cuesta, C.E., Navarro, E. Zdun, U.: Synergies of System-of-systems and Microservices Architectures. In: Proceedings of the International Colloquium on Software-intensive Systems-of-Systems at 10th European Conference on Software Architecture - SiSoS@ECSA 2016, vol. 3 (2016). https://doi.org/10.1145/3175731.3176176. Accessed 13 Jun. 2022
6. Fractal Architecture for IT – FractArc.IT. https://fractarc.it/. Accessed 13 Jun. 2022
7. Warnecke, H.-J. : The Fractal Factory — an Integrating Approach. The Fractal Company, pp. 137–217 (1993). https://doi.org/10.1007/978-3-642-78124-7_4. Accessed 18 Jun. 2022
8. Stecjuka, J., Kirikova, M., Asnina, E.: Fractal modeling approach for supporting business process flexibility. In: Lecture Notes in Business Information Processing, vol. 15 LNBIP, pp. 98–110 (2008). https://doi.org/10.1007/978-3-540-89218-2_8/COVER/. Accessed 13 Jun. 2022
9. Taibi, D., Lenarduzzi, V., Pahl, C.: Architectural patterns for microservices: a systematic mapping study. In: Closer 2018 - Proceedings of the 8th International Conference on Cloud Computing and Services Science, January 2018. https://doi.org/10.5220/0006798302210232
10. Cerny, T., Donahoo, M.J., Trnka, M.: Contextual understanding of microservice architecture. ACM SIGAPP Appl. Comput. Rev. 17(4), 29–45 (2018). https://doi.org/10.1145/3183628. 3183631. Accessed 13 Jun. 2022
11. Santana, C., Alencar, B., Prazeres, C.: Microservices: a mapping study for internet of things solutions. In: NCA 2018 - 2018 IEEE 17th International Symposium on Network Computing and Applications, November 2018. https://doi.org/10.1109/NCA.2018.8548331
12. Krylovskiy, A., Jahn, M., Patti, E.: Designing a smart city internet of things platform with microservice architecture. In: Proceedings - 2015 International Conference on Future Internet of Things and Cloud, FiCloud 2015 and 2015 International Conference on Open and Big Data, OBD 2015, pp. 25–30, October 2015. https://doi.org/10.1109/FICLOUD.2015.55
13. Horsmanheimo, S., Tuomimäki, L., Sanchez, R.R., Andrén, F.P., Andersen, C.A.: ICT requirements in a smart grid environment. In: TSO-DSO Interactions and Ancillary Services in Electricity Transmission and Distribution Networks: Modeling, Analysis and Case-Studies, pp. 61–91, October 2019. https://doi.org/10.1007/978-3-030-29203-4_4/COVER/. Accessed 19 Jun. 2022
14. Hassan, S., Bahsoon, R., Buyya, R. : Systematic scalability analysis for microservices granularity adaptation design decisions. Softw. Pract. Experience, 52(6), 1378–1401, June 2022. https://doi.org/10.1002/SPE.3069. Accessed 13 Jun. 2022
15. Carrasco, A., van Bladel, B., Demeyer, S.: Migrating towards microservices: migration and architecture smells. In: IWoR 2018 - Proceedings of the 2nd International Workshop on Refactoring, co-located with ASE 2018, pp. 1–6, September 2018. https://doi.org/10.1145/3242163.3242164. Accessed 13 Jun. 2022
16. Stranner, H., Strobl, S., Bernhart, M., Grechenig, T.: Microservice Decomposition: A Case Study of a Large Industrial Software Migration in the Automotive Industry (2020). https://doi.org/10.5220/0009564604980505

17. di Francesco, P., Malavolta, I., Lago, P.: Research on architecting microservices: trends, focus, and potential for industrial adoption. In: Proceedings - 2017 IEEE International Conference on Software Architecture, ICSA 2017, pp. 21–30, May 2017. https://doi.org/10.1109/ICSA.2017.24

18. Mikkelsen, A., Grønli, T.-M., Tamburri, D.A., Kazman, R.: Architectural principles for autonomous microservices. In: Hawaii International Conference on System Sciences 2020 (HICSS-53), January 2020. https://aisel.aisnet.org/hicss-53/st/self-adaptive_systems/4. Accessed 13 Jun. 2022

19. Guidelines for performing Systematic Literature Reviews in Software Engineering (2007)

20. Premchand, A., Choudhry, A.: Architecture simplification at large institutions using micro services. In: Proceedings of the 2018 International Conference on Communication, Computing and Internet of Things, IC3IoT 2018, pp. 30–35, March 2019. https://doi.org/10.1109/IC3IOT.2018.8668173

21. Flissi, A., Dubus, J., Dolet, N., Merle, P.: Deploying on the grid with DeployWare. In: Proceedings CCGRID 2008 - 8th IEEE International Symposium on Cluster Computing and the Grid, pp. 177–184 (2008). https://doi.org/10.1109/CCGRID.2008.59

22. Zdun, U., Navarro, E., Leymann, F.: Ensuring and assessing architecture conformance to microservice decomposition patterns. Lecture Notes in Computer Science (including subseries Lecture Notes in Artificial Intelligence and Lecture Notes in Bioinformatics), vol. 10601. LNCS, pp. 411–429 (2017). https://doi.org/10.1007/978-3-319-69035-3_29/COVER/. Accessed 13 Jun. 2022

23. Kirikova, M.: Towards flexible information architecture for fractal information systems. In: Proceedings - International Conference on Information, Process, and Knowledge Management, eKNOW 2009, pp. 135–140 (2009). https://doi.org/10.1109/EKNOW.2009.25

24. Koleini, M., et al.: Fractal: automated application scaling, February 2019. https://arxiv.org/abs/1902.09636v1. Accessed 13 Jun. 2022

25. Visotschnig, M.R., Henke, J., Lucke, D.: A Fractal control system architecture for next generation factories. Procedia CIRP **104**, 1506–1511 (2021). https://doi.org/10.1016/J.PROCIR.2021.11.254

26. Ryu, K., Jung, M.: Agent-based fractal architecture and modelling for developing distributed manufacturing systems, **41**(17), 4233–4255, November 2010. https://doi.org/10.1080/0020754031000149275. Accessed 13 Jun. 2022

27. Bruneton, E., Coupaye, T., Leclercq, M., Quéma, V., Stefani, J.B.: The FRACTAL component model and its support in Java. Softw. Pract. Experience **36**(11–12), 1257–1284, September 2006. https://doi.org/10.1002/spe.767. Accessed 13 Jun. 2022

28. Ryu, K., Son, Y.J., Jung, M.: Framework for fractal-based supply chain management of e-Biz companies. 14(8), 720–733, December 2010. https://doi.org/10.1080/09537280310001647913. Accessed 13 Jun. 2022

29. Canavesio, M.M., Martinez, E.: Enterprise modeling of a project-oriented fractal company for SMEs networking. Comput. Indus. 58(8–9) (2007). https://doi.org/10.1016/j.compind.2007.02.005

30. Zhao, Y., Wu, J., Shu, H.: The fractal management of SOA-based services integration. In: Proceedings of the International Conference on Information Management, Innovation Management and Industrial Engineering, ICIII 2008, vol. 3, pp. 420–424 (2008). https://doi.org/10.1109/ICIII.2008.171

31. Peralta, M.E., Soltero, V.M.: Analysis of fractal manufacturing systems framework towards industry 4.0. J. Manufact. Syst. **57** (2020)

32. Xu, D., Zhao, L., Yao, Y.: Fractal and mobile agent-based inter-enterprise quality tracking and control*. In: Proceedings of the IEEE International Conference on Industrial Technology (2008). https://doi.org/10.1109/ICIT.2008.4608470

33. Deng, X., Peng, J., Huang, H.: Research on the fractal company modeling based on competence. In: IE and EM 2009 - Proceedings 2009 IEEE 16th International Conference on Industrial Engineering and Engineering Management, pp. 2136–2140 (2009). https://doi.org/10.1109/ICIEEM.2009.5344226
34. Asnina, E., Osis, J., Kirikova, M.: Design of fractal-based systems within MDA : platform independent modelling. In: Sigsand-Europe : the 3rd AIS SIGSAND European Symposium on Analysis (2008)
35. Sandkuhl, K., Kirikova, M.: Analysing enterprise models from a fractal organisation perspective - Potentials and limitations. Lecture Notes in Business Information Processing, vol. 92. LNBIP, pp. 193–207 (2011). https://doi.org/10.1007/978-3-642-24849-8_15/COVER/. Accessed 20 Jun. 2022
36. Cheng, X., Qian, J.: Research on supply chain management based on fractal theory. Int. J. Sci. 2(5) (2015)
37. Wu, W., Lu, J., Zhang, H.: Smart factory reference architecture based on CPS fractal. IFAC-PapersOnLine 52(13), 2776–2781 (2019). https://doi.org/10.1016/j.ifacol.2019.11.628
38. Leego, S., Bider, I.: Using fractal enterprise model in technology-driven organisational change projects: a case of a water utility company. In: Proceedings - 2021 IEEE 23rd Conference on Business Informatics, CBI 2021 - Main Papers, vol. 2 (2021). https://doi.org/10.1109/CBI 52690.2021.10061
39. Ahmed, N.: Inspiring a Fractal Approach in Higher Education Institutes' Information Systems in Kurdistan: A Review A Review of most Recent Lung Cancer Detection Techniques using Machine Learning View project Technical SCIENCE View project. https://www.researchg ate.net/publication/344258464. Accessed 20 Jun. 2022
40. Ryu, K., Moon, I., Oh, S., Jung, M.: A fractal echelon approach for inventory management in supply chain networks. Int. J. Prod. Econ. 143(2) (2013). https://doi.org/10.1016/j.ijpe.2012.01.002
41. Herghiligiu, I.V., Lupu, M.L., Robledo, C.: Necessity of change environmental management system architecture - introduction. Quality - Access to Success 13(SUPPL), 5 (2012)

Blockchain Technology Perception in Supporting the Digital Transformation of Supply Chain Management: A Preliminary Study

Franz Nazet[(✉)] [iD] and Michał Kuciapski [iD]

Department of Business Informatics, University of Gdansk, Jana Bażyńskiego 8,
80-309 Gdańsk, Poland
franz.nazet@gmail.com, m.kuciapski@ug.edu.pl

Abstract. The globalization of the world economy increased the role of value chains, where disruptions in the realization of the logistics processes occur repeatedly. Pandemics, military conflicts, and climate change further increase the difficulties in matching demand and supply. Therefore, proper Supply Chain Management (SCM) is crucial and the digital transformation of SCM is considered an important solution for several SCM concerns. Extensive subject matter literature research highlighted that several technologies have been studied and implemented as SCM solutions contrary to the blockchain. As a result, the research aim of the article is to examine the current and future challenges of SCM with particular emphasis on blockchain technology perception in this context. Proper research has been conducted with the use of structured in-depth interviews with SCM experts. Pilot study results highlighted that blockchain is perceived as a technology with significant potential to support the creation of Digital Supply Chains. They also pointed out that blockchain is not the technology of the first choice when planning the digital transformation of SCM.

Keywords: Blockchain · Supply Chain Management · Digital transformation · COVID-19 · Climate change

1 Introduction

Supply Chain Management (SCM) is a comprehensive concept for planning, controlling, and integrating corporate activities along the value chain [22]. SCM creates a network of all cooperating areas, referred to as the supplier-manufacturer-customer network. The key goals of SCM are mostly related to a significant increase in cross-process efficiency and are based on economic and ecological motivation [27].

The globalization of the world economy has made the world to be more integrated and connected. As a result of the ever more widely branched value chains, disruptions in the realization of the process occur repeatedly [12]. Dispersion of supply chains carries several challenges in conjunction with pandemics, wars, and climate change.

At the beginning of the COVID-19 pandemic, many sectors were connected through a complex network of supply chains and logistics, but hardly any activities were evidenced

J. Maślankowski et al. (Eds.): PLAIS EuroSymposium 2022, LNBIP 465, pp. 126–137, 2022.
https://doi.org/10.1007/978-3-031-23012-7_8

during the COVID-19 pandemic [12]. Due to the strict lockdown, manufacturing, and logistics activities have been suspended, affecting the demand and supply of various products because of restrictions imposed on shopkeepers and retailers. Difficulties have increased in matching supply and demand in a vast majority of networks because of changing scenarios with the growth of infected cases and recovery. The World Health Organization (WHO) has recommended social distancing to control the spread among the public with the imposition of necessary lockdowns. Most of the regular flights were canceled, so air freight was no longer available for the global supply chain. A quick switch from air freight to sea or train freight was needed and extended the shipping time from 2 weeks to 12 weeks [25]. Those disruptions of normal material supply and semi-finished goods significantly disturbed the normal production of goods.

Also, military conflicts like Russia's invasion of Ukraine significantly negatively influence SCM. The impact has been felt around the world as Russia and Ukraine are major commodity producers [23]. The disruptions have caused global prices to increase rapidly, especially for natural gas and oil. Food costs have also soared with wheat which Russia and Ukraine account for 30 percent of global exports, reaching record levels [6]. Higher prices for commodities such as food and energy continued to drive up inflation, reducing the value of incomes and weighing on demand. Neighbouring economies faced disruption in trade, supply chains, and remittances. The higher energy price increased the freight and shipping costs, also manufacturing processes that are very energy intensive, such as plastics processing, are becoming increasingly expensive. Business confidence and greater investor uncertainty have weighed on asset prices, potentially leading to capital outflows from emerging markets [14].

Climate change is also having an important impact on SCM. Two main trends in climate change can be distinguished. The first one involves addressing carbon management issues, with most companies trying to minimize their carbon footprint across their supply chain [26]. The second process involves managing physical impacts, such as temperature fluctuations, extreme weather patterns, very low humidity, etc. [15]. Climate-sensitive businesses such as agriculture or food production and supply could be affected on a large scale. Heavy rainfall may lead to floods, resulting in suspension of navigation, and can cause damage to the infrastructure on waterways. Long periods of drought may lead to low water levels, limiting the cargo-carrying capacity of vessels and increasing transport costs [20].

Regarding indicated SCM threats, it is necessary to have as much information as possible about the status of the respective actuators in the supply chain, to be able to react quickly to the challenges regarding pandemics, military conflicts, and weather anomalies. Digital transformation of Supply Chain Management is considered an important solution for many SCM concerns where studies resulting from the innovations of SCM should limit negative impacts connected with disruptions of the supply chains [18]. Digital Supply Chains (DSC) are defined as the development of information systems and the adoption of innovative technologies that strengthen the supply chain's integration and agility and thus improve the organization's customer service and sustainable performance [1].

Different technologies and solutions are available to gather and process data connected with the supply chain. The phenomenon of the DSC is in its preliminary steps starting from Büyüközkan and Göçer in 2018 [7], and its potential is still relatively unknown. Conducted research in this context mainly focused on the concept of industry 4.0 and the technological drivers of the DSC, such as the Internet of Things (IoT), big data, and cloud computing.

Blockchain is an important technology nearly not studied for the digital transformation of SCM. Such a belief is confirmed by preliminary analysis of related research searched databases, such as ACM Digital Library, IEEE Xplore, Springer, Web of Science, Scopus, Science Direct, AIS digital library as well as the EBSCOhost multi-source. In total, 52 journals and 133 conference proceedings that are directly or indirectly related to the keywords such as Supply Chain Management, Digital Supply Chain, blockchain, and digital transformation were explored. Proper studies of blockchain implementation in SCM might address some of SCM's limitations and challenges.

Blockchain is perceived as a revolutionary technology that has a major impact on modern society due to its transparency, decentralization, and security features [5]. Blockchain has been successful in its first application of cryptocurrencies, such as Bitcoin. Recently, academics, industrialists, and researchers have been actively exploring various aspects of blockchain as an emerging technology [5]. Since blockchain can exchange data securely in a distributed fashion, it is starting to affect how organizations are managed, supply chain relationships are structured, and transactions are executed [21]. Combined with other technologies such as the Internet of Things, blockchain can be used to create durable, sharable, and actionable records of every moment a product is in its supply chain, thereby increasing the efficiency of the global economy. This helps improve products' traceability, authenticity, and legality [2].

The research aim of the article is to examine the current and future challenges and limitations of digital transformation of Supply Chain Management with particular emphasis on blockchain technology perception in this context. Therefore, it will allow identifying decision-makers perceptions of blockchain technology in addressing the challenges of digital transformation of Supply Chain Management. This is the initial step of a broader study to create a theoretical framework for blockchain implementation for the digital transformation of SCM.

2   Related Research

Subject matter literature review with the use of databases, such as ACM Digital Library, IEEE Xplore, Springer, Web of Science, Scopus, Science Direct, AIS digital library, Google Scholar as well as E-Theses Online Service (EThOS), Social Science Research Network (SSRN), and Public Library of Science (PLOS) was conducted with the keywords as blockchain, Supply Chain Management / SCM, Digital Supply Chain / DSC, blockchain, and digital transformation, and Value Chain. Many articles from journals and conference proceedings have been studied and allowed to explore blockchain technology utilization in various sectors. As a result, Table 1 contains a synthesis of 8 articles chosen according to the significance of the study results.

The findings juxtaposed in Table 1 point out that blockchain technology has a wide range of applications in various business processes, also supporting their digital transformation. Unfortunately, Supply Chain Management is a missing gap. Table 1 supports

Table 1. Usage of blockchain technology

Article	Sector	Finding
[13]	Retail	Blockchain has demonstrated its potential for providing greater transparency, veracity, and trust in food information for a small area. Ensuring value for all participants in the ecosystem will be critical to wider adoption
[10]	Energy	Blockchain-based applications can integrate Internet of Things devices in the power grid to manage the e-mobility infrastructure, automate billing and direct payment, and issue certificates regarding the origin of electricity
[27]	Real estate	Real estate sector transformation requires mechanisms for adopting smart contracts, where every key stakeholder knows the other parties, how much they pay, the conditions, etc. The 'anonymity' of transactions is crucial and might be achieved with the use of blockchain
[16]	Automotive	The automotive value chain comprises a complex manufacturing network encompassing several departments, companies, and even countries. A blockchain-based preliminary theoretical model was proposed to prevent safety and quality deficiencies in manufacturing processes
[11]	Agriculture	Blockchain technology can play a key role in reducing intermediaries, allowing small farmers to connect directly with customers and end customers, reducing logistic costs, and preventing corruption during contractors' selection
[19]	Transport	The digital transformation of communication is slowly ongoing because of the manager's resistance. Paper documents will be exchanged between the stakeholders in most transport transactions, even the electronic exchange of documents with the support of blockchain can speed up those processes
[4]	Production	Blockchain technology can be used for production scheduling and planning in production. Multi-criteria optimization methods to build consensus on the problems of production and human resources planning are still missing
[3]	Finance	It would be beneficial to integrate financial transactions using blockchain technology in ERP systems. Moreover, two constructs, such as regulatory support and experience that encourage customers' trust in blockchain-based applications. At a certain level of experience, users feel trusted using blockchain-based finance applications

the study results of Lim et al. [17] that blockchain technology is nearly not used in Supply Chain Management. There is currently only one, and pilot study by IBM and Walmart using blockchain technology to track food chains, specifically pork and mango [13]. The study is only a starting point to apply it to other areas of the food chain, and if the implementation is successful, it might be possible to extend it to other related areas.

3   Research Method

The research data collection was obtained with a structured in-depth interview method. Since the study topic is highly complex and requires respondents' broad interdisciplinary knowledge of economy, management, and technology, the elaborated survey was conducted among subject matter experts as particular case studies. A more controlled structured interview uses listings to explore the meaning of terms and the rules governing them. In-depth interviews can provide rich and in-depth information about the experiences of individuals [9].

To study as wide a range of Supply Chain Management concerns, challenges, and digital transformation technologies, four experts from different hierarchical levels and positions with extensive experience in SCM also responsible for the digitalization of SCM processes were interviewed (Table 2).

Table 2. Surveyed experts

Number	Country	Title	Department	Area of Expertise
1	Germany	Senior Expert	Strategic Planning	SCM Planning SCM Design
2	Germany	Team leader	Operational Procurement	Electronic Manufacturing
3	Germany	Head of SCM	Supply Chain Management	Functional Shoe Manufacturing
4	Germany	Team leader	Work scheduling	Special machine construction

Structured in-depth interviews with experts were based on the elaborated survey. The survey consisted of 2 sections:

- Classification data - this section was split into two subsections "Company data" and "Personal data" consisting of 10 questions. The company data subsection provided information about the company. The personal data subsection provided information about experts' experience in logistics, especially the digitalization of SCM.
- Supply Chain Management concerns, challenges, and limitations - this section is divided into three subsections "Digital Transformation," "Information exchange," and "Business models" and includes 33 questions in total.

Depending on the question type two scales were used 5-Point Likert scale and 3-PoinLikert. The 3-Point Likert scale had two versions:

- 0 = not applicable / I do not know, 1 = not implemented, 2 = implementation ongoing, 3 = implemented
- 0 = not applicable / I do not know, 1 = not planed, 2 = basic plans, 3 = detailed plans

Following the Likert scale used different formulas for measuring results were utilized:

- 5-Point Likert scale - the sum of results / number of respondents,
- 3-Point Likert scale - number of responses with values of 2 or 3 / number of respondents.

6 from 33 questions of the "Supply Chain Management concerns, challenges, and limitations" section related to the current study (Table 3). The other ones regarded further steps of a broader study. The questionnaire was constructed to get a complete overview of the limitations from current and future perspectives in the field of Supply Chain Management. Consideration was given to the potential of technologies in a deeper examination.

Table 3. Interview questions

Number	Question	Section
4.5	What limitations exist with the exchange of information with your business partners and public institutions in the context of Supply Chain Management?	Information Exchange
4.6	What limitations should exist with the exchange of information with your business partners and public institutions in the context of Supply Chain Management in the future (next 10 years)?	Information Exchange
3.1	How would you rate the potential of the listed below technologies in supporting Supply Chain Management?	Digital Transformation
3.7a	How would you rate the difficulty of implementing technologies that may support Supply Chain Management?	Digital Transformation
3.3	What innovative technologies have you introduced during the last 5 years for Supply Chain Management?	Digital Transformation
3*	What innovative technologies do you plan to implement during the next 5 years for Supply Chain Management?	Digital Transformation

Question 3* (Table 3) was not part of the original survey and was introduced later as a new question during the complementary interview stage to extend and standardize the structure of all aspects for current and future perspectives. For particular questions, multiple choice answers from a dozen to several dozen were proposed as a result of an extensive subject matter literature review dedicated to SCM and DSC. Experts were also always able to extend the list.

The structured in-depth interviews were conducted as part of a process consisting of three stages:

1. Test – was conducted in April 2021 with one expert to validate the survey structure and receive feedback. After the test, some changes to the questionnaire were introduced, such as extending answer lists.

2. Pilot - the pilot of the survey was taken in June 2021 with two experts to assess the logical nature of the responses received. Once again minor changes were required, like questions and especially answers formatting.
3. Regular – 4 structured interviews with CAWI (Computer-Assisted Web Interview) method were conducted from September 2021 to May 2022. Half of the interviews were conducted online via Microsoft Teams, while the other half were realized on-site. The online tool SurveyMonkey was used to collect data. During the face-to-face or online interviews, the experts could ask for a detailed explanation of questions or answers.

During the evaluation of the collected data, it was determined that additional data must be collected to obtain the complete picture, and to extend and standardize the structure of all answers for current and future perspectives. The complementary interview was elaborated as an extension of the initial one. The structured in-depth interviews for the complementary interview stage were conducted during two sub-stages:

1. Pilot - was taken on June 2022 with two experts to validate the survey structure and assess the logical nature of the responses received. No changes had to be introduced.
2. Regular - the remaining two structured interviews were conducted until August 2022. All the interviews were conducted online via Microsoft Teams and SurveyMonkey to collect data.

4 Research Results

In general, three experts had more than 15 years of experience, and one had 6 to 10 years of experience. In Supply Chain Management, one expert had 11 to 15 years of experience, one between 6 and 10 years of experience, and two had 3 to 5 years of experience. This confirms that respondents could be treated as subject matter experts.

Table 4 presents structured in-depth interview results regarding the challenges of Supply Challenges Management from current and future perspectives respectively. Challenges ratings determined on a scale of 1 to 5 are sorted by the current period – left column in Table 4. The last column contains the predicted values for the future, that is in 10 years.

According to Table 4, several challenges for SCM exist. Importantly, few of them have a high ranking, that is over three. Such ones are high costs and transparency in supply chains. It is noticeable, that experts perceive challenges of digital transformation to be more important in the future perspective than they are currently (Table 3). The ranking structure also changes significantly, but still many similarities with the current period exist.

Table 4 results highlight that there is a need to explore technologies and develop related solutions that will support SCM digital transformation. Therefore, as a part of the study as presented in Table 5 experts rated technologies' usability in the digital transformation of Supply Chain Management digital transformation. Also, they assessed the difficulty of their implementation for SCM digital transformation, marking which technologies they have implemented or planned to.

Table 4. Challenges of Supply Chain Management

Current	Challenge	Future
3,5	High costs	2
3,2	Transparency in supply chains	3,75
2,75	Decentralization of workplaces, production sites, or systems	3,5
2,75	Increase in risks and disruptions	2,25
2,75	Networking and collaboration between companies	3,25
2,75	Traceability of components and products	4
2,5	Government regulation/compliance	2
2,5	Sustainability	2,25
2,25	Business analytics	3,25
2,25	Complexity of operations	2,75
2	Digitalization of business processes	1,75
2	Individualization of customer expectations	2,25
2	Lack of qualified personnel	2
1,75	Automation of processes or operations	3,25
1,75	Customs processing	2
1,5	Detailed planning is too difficult	1,5
1,25	Fluctuating customer demand	1,5
1	Changing buyer behavior (e-commerce)	1,25

Table 5. Technologies supporting the digital transformation of Supply Chain Management

Potential Difficulty	Technology	Implemented Planed
4,75	Mobile technologies	75%
3		100%
4,5	Artificial intelligence	0%
1,5		25%
4,5	Intelligent process automation	25%
2		75%
4,25	RFID tracking management	50%
1		100%
4	3D Printing for components	50%

(continued)

Table 5. (*continued*)

Potential Difficulty	Technology	Implemented Planed
1,75		75%
4	Cloud computing	50%
1,75		100%
4	Robotics and robot process automation	25%
3,5		75%
3,75	5G	0%
1,5		25%
3,75	Big data	50%
2		25%
3,25	Cybersecurity	25%
1,25		75%
3,25	Drones	0%
1,25		0%
3,25	Internet of Things	0%
0,25		75%
3	Virtual reality solutions	25%
1		25%
2,75	Agent system	0%
1,5		50%
2,5	**Blockchain**	0%
0		0%
2	DevOps	0%
0		25%
1,75	Augmented reality	25%
0,75		25%

According to Table 5, many technologies support the digital transformation of SCM. Importantly, blockchain with a rating of 2,5, despite being perceived by the experts as moderately useful for developing Digital Supply Chains is assessed as one of the lowest importance for the digital transformation of SCM.

Study results presented in Table 5 highlight great differentiation in the levels of implementation or plan to implement individual technologies in companies represented by experts. Moreover, many technologies have not been implemented in companies in which experts are employed, where blockchain is among them.

5 Discussion

Research results included in Table 4 highlight that there are a lot of concerns connected with Supply Chain Management. Importantly, their perceived by experts significance increases for the future period – during 10 years perspective. This is consistent with the other studies highlighting the increase in the future negative impact of factors such as pandemics [12, 25], military conflicts [6, 23], and climate change [8, 15, 26] on SCM.

Therefore, the requirement for the digital transformation of SCM will increase what is consistent with study results presented in Table 5. Even though most of the surveyed experts have implemented in their organization several technologies to create Digital Supply Chains they are still planning further extensive activities in this area. Blockchain together with drones is an exception. None of the experts implemented it for the creation of DSC, and neither plans to use it for the digital transformation of SCM in the future.

Blockchain, despite not being utilized for the digital transformation of SCM cannot be defined as unuseful technology for this purpose. According to the results presented in Table 5, with a rating of 2,5, it is perceived by experts as a technology that should support the digital transformation of Supply Chain Management.

There is one key reason explaining why blockchain perceived by experts as a useful technology for the digital transformation of SCM is neither used nor planned to be applied in the development of DSC. Only initial studies were conducted showing how to use blockchain technology in the field of Supply Chain Management. The pilot study of Walmart and IBM shows that there is a possibility to use blockchain technology to improve traceability. However, in the pilot study, it is used just for two products [26]. Therefore, the lack of blockchain-based solutions, guidelines, and successful implementation stories makes it impossible to apply blockchain as a technology for the digital transformation of SCM.

Study results included in Tables 4 and 5 allow for a few implications for the theory. First, there is a role for blockchain technology in the field of Supply Chain Management. Despite blockchain is not a mainstream solution, it might be an important one for solving particular highly rated SCM concerns such as transparency in supply chains (Table 4). Therefore, broad research should be conducted to discover its potential and, more importantly, to elaborate theoretical frameworks and models as a starting point for the development of practical IT solutions.

Research results also highlight several important practical implications. First, survey results among experts point out there is a significant necessity for blockchain solutions (Table 4) that will support the creation of Digital Supply Chains. Especially startups should initiate the design of innovative blockchain-based solutions supporting the optimalization of Supply Chain Management. Second, as blockchain usability for the digital transformation of SCM was rated as moderate, decision-makers if initiating the creation of DSC should concentrate on the use of other higher-rated technologies. Such ones are mobile technologies, artificial intelligence, intelligent process automation, RFID tracking management, 3D printing for components, cloud computing, robotics, and robot process automation (Table 4). On the other side they should not underestimate blockchain and for already existing mature digital transformations of SCM, regularly check whether new blockchain-based solutions exist, that might optimize current SCM systems.

The pilot study's main limitation is survey research among experts from one country. Therefore, the number of experts will be increased with new ones from other countries, allowing research results to include a more global perspective. Moreover, further research aim is to create a complex theoretical framework for blockchain implementation for the digital transformation of SCM.

6 Conclusion

The authors investigated blockchain technology perception for supporting the digital transformation of Supply Chain Management (SCM), where the research aim to rate technologies usability in SCM digital transformation was achieved. The primary outcome of the performed study is that blockchain is perceived by experts as a technology that should significantly support the digital transformation of Supply Chain Management. However, it is considered a useful, but not essential technology during the development of Digital Supply Chains. Despite its potential, blockchain is hardly used at all during the digital transformation of SCM. The key reason of such a situation is that despite the expert's willingness to implement blockchain for digitization of SCM, available solutions are at h the stage of pilot studies.

Study results contribute to both theory and practice. The key implication for theory is that blockchain technology adoption in SCM digital transformation studies should be started on a broad scale to elaborate theories, models, and frameworks. The crucial implication for practice is to adopt other technologies like the Internet of Things or Cloud Computers for the digital transformation of Supply Chain Management as blockchain solutions in this context are not available. On the other side mature Digital Supply Chains could be optimized if proper blockchain solutions were elaborated.

References

1. Ageron, B., Bentahar, O., Gunasekaran, A.: Digital supply chain: challenges and future directions. Supply Chain Forum Int. J. **21**, 133–138 (2020)
2. Alabbasi, Y.: Governance and legal framework of Blockchain technology as a digital economic finance. Int. J. Innov. Digital Econ. **11**, 52–62 (2020)
3. Albayati, H., Kim, S.K., Rho, J.J.: Accepting financial transactions using Blockchain technology and cryptocurrency: a customer perspective approach. Technol. Soc. **62**, 101320 (2020)
4. Balon, B., Kalinowski, K., Paprocka, I.: Application of Blockchain technology in production scheduling and management of human resources competencies. Sensors (Basel, Switzerland) **22** (2022)
5. Bhutta, M.N.M., et al.: A survey on Blockchain technology: evolution. architecture and security. IEEE Access **9**, 61048–61073 (2021)
6. Bluszcz, J., Valente, M.: The economic costs of hybrid wars: the case of ukraine. Defence Peace Econ. **33**, 1–25 (2022)
7. Büyüközkan, G., Göçer, F.: Digital supply chain: literature review and a proposed framework for future research. Comput. Ind. **97**, 157–177 (2018)
8. Dasaklis, T.K., Pappis, C.P.: Supply chain management in view of climate change: an overview of possible impacts and the road ahead. JIEM **6** (2013)

9. Dicicco-Bloom, B., Crabtree, B.F.: The qualitative research interview. Med. Educ. **40**, 314–321 (2006)
10. Höhne, S., Tiberius, V.: Powered by Blockchain: forecasting blockchain use in the electricity market. IJESM **14**, 1221–1238 (2020)
11. Hrustek, L.: Sustainability driven by agriculture through digital transformation. Sustainability **12**, 8596 (2020)
12. Illahi, U., Mir, M.S.: Maintaining efficient logistics and supply chain management operations during and after coronavirus (COVID-19) pandemic: learning from the past experiences. Environ. Dev. Sustain. **23**(8), 11157–11178 (2021). https://doi.org/10.1007/s10668-020-011 15-z
13. Kamath, R.: Food traceability on Blockchain: Walmart's pork and mango pilots with IBM. JBBA **1**, 1–12 (2018)
14. Kammer, A., Azour, J., Aemro, A.S., Goldfajn, I., Rhee, C.: How War in Ukraine is Reverberating Across World's Regions. https://blogs.imf.org/2022/03/15/how-war-in-ukraine-is-reverberating-across-worlds-regions/. Accessed 18 Aug 2022
15. Karthick, S., Kermanshachi, S., Rouhanizadeh, B., Namian, M.: Short- and Long-Term Health Challenges of Transportation Workforce due to Extreme Weather Conditions, pp. 39–51 (2021)
16. Kuhn, M., Funk, F., Franke, J.: Blockchain architecture for automotive traceability. Procedia CIRP **97**, 390–395 (2021)
17. Lim, M.K., Li, Y., Wang, C., Tseng, M.-L.: A literature review of Blockchain technology applications in supply chains: a comprehensive analysis of themes, methodologies and industries. Comput. Ind. Eng. **154**, 107133 (2021)
18. Liu, C.: Risk prediction of digital transformation of manufacturing supply chain based on principal component analysis and backpropagation artificial neural network. Alex. Eng. J. **61**, 775–784 (2022)
19. Malyavkina, L.I., Savina, A.G., Parshutina, I.G.: Blockchain technology as the basis for digital transformation of the supply chain management system: benefits and implementation challenges. In: 1st International Scientific Conference, "Modern Management Trends and the Digital Economy: From Regional Development to Global Economic Growth", pp. 10–15. Atlantis Press (2019)
20. Maternová, A., Materna, M., Dávid, A.: Revealing causal factors influencing sustainable and safe navigation in central Europe. Sustainability **14**, 2231 (2022)
21. Mehta, D., Tanwar, S., Bodkhe, U., Shukla, A., Kumar, N.: Blockchain-based royalty contract transactions scheme for Industry 4.0 supply-chain management. Inf. Process. Manag. **58**, 102586 (2021)
22. Mentzer, J.T., et al.: Defining supply chain management. J. Bus. Logist. **22**(2), 1–25 (2019)
23. Orhan, E.: The effects of the Russia-Ukraine war on global trade. J. Int. Trade Logistics Law **8**(1), 141–146 (2022)
24. Peck, H.: Drivers of supply chain vulnerability: an integrated framework. Int. J. Phys. Distrib. Logist. Manag. **35**, 210–232 (2005)
25. Singh, S., Kumar, R., Panchal, R., Tiwari, M.K.: Impact of COVID-19 on logistics systems and disruptions in food supply chain. Int. J. Prod. Res. **59**, 1993–2008 (2021)
26. Sodhi, M.S., Tang, C.S.: Supply chain management for extreme conditions: research opportunities. J Supply Chain Manag **57**, 7–16 (2021)
27. Ullah, F., Al-Turjman, F.: A conceptual framework for Blockchain smart contract adoption to manage real estate deals in smart cities. Neural Comput. Appl. (2021)
28. Wieland, A.: Dancing the supply chain: toward transformative supply chain management. J. Supply Chain Manag. **57**(1), 58–73 (2021)

Author Index

Printed in the United States
by Baker & Taylor Publisher Services